Fundamentals of Plant Biochemistry & Biotechnology

THE AUTHORS

Mr. Vikas Singh, He is pursuing his Ph.D. Agricultural Biotechnology from SVPUA&T Meerut, UP, India. He completed his M.Sc. Agricultural biotechnology in 2019 from NDUA&T Kumarganj Ayodhya,UP, India. He got 4th rank in UPCATET for his Ph.D. in 2021. He has published 9 book chapters and 12 Hindi articles in different Magazines.

Prof. Mukesh Kumar is working as an Associate Professor in the Department of Biotechnology, S V P University of Agriculture and Technology, Meerut (UP). His specialization is in Genetic Engineering for Stress Management and Plant Tissue Culture. He has more than nineteen years of experience in Teaching and Research in Graduate and Post Graduate Classes. The author published more than fifty research papers in reputed national and international journals.

Prof. Rakesh Singh Sengar is Professor and Head, Department of Biotechnology, College of Agriculture in Sardar Vallabhbhai Patel University of Agriculture & Technology, Meerut. Dr. Sengar has about 29 years of professional experience in different aspects of educational planning, administration, institution building, teaching, research and extension experience to his credit. He has supervised 16 Ph.D. students and published more than 166 research papers in India and foreign journals of repute. He has published 972 popular articles in Hindi and English languages in important journals and magazines. He has also published 18 books for scientific community and farmers. Dr. Sengar has delivered more than 225 talks at All India Radio and Television at Delhi, Lucknow, Rampur and Bareilly centers.

Fundamentals of Plant Biochemistry & Biotechnology

– Authors –

Mr. Vikas Singh

Division of Plant Biotechnology
College of Biotechnology
Sardar Vallabhbhai Patel University of Agriculture and technology,
Meerut 250110 (U.P)

Prof. Mukesh Kumar

Division of Plant Biotechnology
College of Biotechnology
Sardar Vallabhbhai Patel University of Agriculture and technology,
Meerut 250110 (U.P)

Prof. Rakesh Singh Sengar

Head

Division of Plant Biotechnology
College of Biotechnology
Sardar Vallabhbhai Patel University of Agriculture and technology,
Meerut 250110 (U.P)

2024

Daya Publishing House®
A Division of
Astral International Pvt. Ltd.
New Delhi – 110 002

ISBN: 9789359194264

Published by : **Daya Publishing House®**
A Division of
Astral International Pvt. Ltd.
– ISO 9001:2015 Certified Company –
47636/23, Ansari Road, Darya Ganj
New Delhi-110 002
Ph. 011-43549197, 8130496929
e-mail: info@astralint.com
Website: www.astralint.com

"This Book is Dedicated

To My Beloved Parents

& Elder Brother"

SARDAR VALLABHBHAI PATEL UNIVERSITY OF AGRI.& TECH.,MEERUT - 250110 (U.P.) INDIA

सरदार वल्लभभाई पटेल कृषि एवं प्रौद्योगिकी विश्वविद्यालय मेरठ - 250110 (उ.प्र.), भारत

डॉ. के. के. सिंह
कुलपति
Dr. K. K. Singh
Vice Chancellor

Phone: 0121-2888522(O)
: 0121-2888566 (R)
Web: svbpmeerut.ac.in
E-mail : vc@svpuat.edu.in

Foreword

Dr. K. K. Singh
Vice Chancellor

It is with great pleasure and enthusiasm that I write this foreword for the newly proposed textbooks on the fundamentals of plant biochemistry and biotechnology. As we witness remarkable advancements in the field of plant sciences, these books stand as a testament to the ever-evolving nature of our understanding of plants and their potential for solving global challenges. It is a very useful book based on the ICAR syllabus. This book can be used for each and every student of UG and PG mostly. The idea behind this book was students' doubts can be simplified in a simple way. The design text has been in a simple way and helps for read and learn also. It will be helping the students gain knowledge of their own subject and clear doubts as well.

These textbooks, I hope, will find an appropriate position in educational courses, libraries and lead to further advances in our understanding and application of plant biochemistry and biotechnology. The authors need to be congratulated for bringing out publication

(K. K. Singh)

Preface

This book has been written for those students of agriculture who are preparing for various entrance enroll in B.Sc.(ag), M.Sc. and Ph.D. biotechnology. This book is good for all competitive exam such as UPCATET, BHU, ICAR NET, ICAR JRF, ICAR SRF and others who related from basic biochemistry, plant tissue culture as well as plant biotechnology subjects. This book is helpful because we focused on concept based and describe in simple way. I hope this book clarify to all doubt in your mind.

This book is based on ICAR syllabus. In which categories of chapters in serial way that will be helps in read to every student. We have conceptual things into the bullet point so that they became easy to memorize for a long time. In this book include multiple choices question (MCQ) and fill in the blanks that are based on syllabus.

This is first edition of book and you will read this book I hope this book will be helps for enhance to the knowledge. After read this book, please give feedback to me.

Authors

Vikas Singh

Mukesh Kumar

Rakesh Singh Sengar

Contents

Syllabus of Fundamentals of Plant Biochemistry and Biotechnology

1. Plant Biochemistry

Introduction: Biochemistry and their importance, Properties of water, pH and Buffer.

Carbohydrates: Importance and Classification. Structure of Monosaccharide, Reducing and oxidizing properties of monosaccharides, Mutarotation; Structure of Disaccharide and Polysaccharides.

Lipid: Importance and classification; Structure and Properties of fatty acids; Storage lipid and membrane lipids.

Protein: Importance of proteins and Classification; Structures, Titration and Zwitterion nature of amino acids; Structural organization of proteins.

Enzymes: General properties; Classification; Mechanism of action; Michalis and Menten, Line Weaver Burk equation and plots; Introduction to allosteric enzymes.

Nucleic acid: Importance and classification; structure of Nucleotide, A, B & Z DNA; RNA: Types and Secondary and Tertiary Structure.

Metabolism of carbohydrates: Glycolysis, TCA cycle, Glyoxylate cycle, Electron Transport Chain.

Metabolism of Lipids: Beta oxidation, Biosynthesis of fatty acids.

2. Concepts and application of plant biotechnology

Tissue Culture: Scope, Organ culture, cell suspension culture, callus culture, anther culture, Pollen culture and ovule culture and their application; Micropropagation method; organogenesis and embryogenesis, Synthesis seed and their significance; Embryo rescue and its significance; somatic hybridization and cybrid; Somaclonal variation and its use in crop improvement; cryo- preservation;

Plant Biotechnology: Introduction to recombinant DNA method: Physical (gene gun method), chemical (PEG mediated) and Agrobacterium mediated gene transfer method. Transgenic and its improvements; PCR Techniques and its application; RFLP, RAPD, SSR; Marker Assisted Breeding in crop improvement; Biotechnology regulations.

Unit-1

Plant Biochemistry

Introduction

Biochemistry: Biochemistry is a scientific discipline that studies the chemical processes and molecules found in living organisms. It combines principles of biology and chemistry to understand the structure, function, and interactions of biological molecules, such as proteins, carbohydrates, lipids, and nucleic acids.

Biochemists investigate various aspects of life at the molecular level, including the metabolic pathways, signaling mechanisms, and genetic information flow within cells. By elucidating the molecular basis of biological processes, biochemistry provides insights into the fundamental mechanisms of life and forms the foundation for advancements in medicine, biotechnology, and other related fields.

Plant Biochemistry: Plant Biochemistry is the study of chemical processes within plants. Some of the processes that are controlling the biochemical signaling process and the flow of chemical energy through metabolism within plants.

Importance

Biochemistry is used to enhance plant growth, yield, and quality as a consequence of optimizing fertilizer components. Crop improvement has also been improved by way of increased tolerance to biotic and abiotic stresses, alongside augmented nutritional value.

Properties of water: Water is a unique and essential substance for life, possessing several distinctive properties:

- **High Polarity:** Water is a polar molecule, meaning it has a partial positive charge on one end (hydrogen) and a partial negative charge on the other end (oxygen).
- **High Specific Heat Capacity:** Water has a high specific heat capacity, meaning it can absorb and store a significant amount of heat energy before its temperature increases.
- **Excellent Solvent**: Water is often called the "universal solvent" because of its ability to dissolve a wide range of substances. Its polarity allows it

to interact with and dissolve other polar and ionic compounds, making it crucial for biological processes such as nutrient transport and metabolic reactions.

- **High Heat of Vaporization**: Water has a high heat of vaporization, meaning it requires a significant amount of heat energy to change from a liquid to a gas (vapor) state.
- **Density Anomaly**: Most substances become denser as they solidify, but water behaves differently. It reaches its maximum density at 4°C, and when it freezes to form ice, it expands, becoming less dense. This unique property allows ice to float on water, insulating the underlying liquid and providing a protective habitat for aquatic life during freezing conditions.
- **High Cohesion and Adhesion**: Water exhibits strong cohesion, meaning water molecules tend to stick together, forming hydrogen bonds. It also displays adhesion, allowing it to adhere to and interact with other surfaces. These properties enable water to be transported against gravity in plants (capillary action) and facilitate the movement of water in small tubes or vessels.
- **Versatile Biological Medium**: Water provides an ideal medium for various biological processes. It participates in chemical reactions, supports cellular structures, acts as a lubricant and cushioning agent, and aids in the transportation of nutrients and waste products within organisms.

Water Properties in a Simple Way

- A water molecule consists of hydrogen and oxygen molecule.
- Maximum water density occurs at 4°(c)
- Heat of fusion is 44 j/mol for water.
- Freezing temperature of water occurs at 0°C.

pH value: pH refers to the measurement of hydrogen ion concentration.

$pH= -\log |H^+|$

Buffer: A buffer is a solution that helps maintain the pH (acidity or alkalinity) of a system by resisting changes in its hydrogen ion concentration ([H+]). It consists of a weak acid and its conjugate base or a weak base and its conjugate acid. Buffers play a crucial role in biological systems, as they help stabilize the pH levels necessary for the proper functioning of enzymes, cellular processes, and overall physiological balance.

Key points about buffers include:

- **Acid-Base Balance**: Buffers work by accepting or donating protons (H+) to maintain the balance between acidic and basic conditions. When an acid is added to a buffer solution, it combines with the conjugate base, minimizing the increase in hydrogen ion concentration and preventing a significant change in pH. Similarly, when a base is added, it reacts with the weak acid component of the buffer, preventing a drastic decrease in pH.

- **Buffering Capacity**: The ability of a buffer to resist changes in pH is known as its buffering capacity. Buffers have the highest buffering capacity at a pH close to their pKa (the negative logarithm of the acid dissociation constant). This is the pH at which the weak acid and its conjugate base are present in equal amounts, maximizing their ability to neutralize added acids or bases.
- **Range of Effectiveness**: Buffers have an optimal pH range where they are most effective. Outside this range, their buffering capacity decreases significantly. For example, a buffer with a pKa of 7.4 is well-suited for maintaining physiological pH in the human body, as this corresponds to the pH of the blood.
- **Common Buffer Systems**: Some common buffer systems include the bicarbonate buffer system in blood, which helps regulate pH, and the phosphate buffer system in cells and biological fluids. Additionally, many laboratory buffers are based on acetic acid/acetate or citric acid/citrate systems.
- **Buffer Preparation**: Buffers can be prepared by mixing a weak acid and its conjugate base or a weak base and its conjugate acid in appropriate proportions. The choice of buffer system depends on the desired pH range and the specific requirements of the experiment or application.

Plant Biotechnology

Plant biotechnology refers to the application of scientific techniques and tools to manipulate and modify plants for various purposes. It involves utilizing knowledge from disciplines such as genetics, molecular biology, and biochemistry to improve plant characteristics, enhance crop yields, and develop new plant varieties with desired traits.

Plant biotechnology encompasses a wide range of techniques and applications, including:

- **Genetic engineering**: This involves the transfer of specific genes or DNA sequences into plant cells to introduce new traits or modify existing ones. It allows scientists to enhance traits like disease resistance, herbicide tolerance, and nutritional content in crops.
- **Tissue culture and micropropagation**: Plant tissue culture involves growing plant cells or tissues in a controlled laboratory environment. It enables the production of large numbers of identical plants from a small tissue sample, which is particularly useful for the rapid propagation of rare or valuable plant species.

- **Plant transformation**: Transformation involves introducing foreign DNA into the plant genome, resulting in genetically modified plants. This technique has been instrumental in developing crops with improved traits, such as insect resistance, increased yield, and enhanced nutritional value.
- **Molecular marker-assisted breeding**: Molecular markers are DNA sequences that can be used to identify and select plants with desired traits during the breeding process. This technique allows for more precise and efficient selection of plants with specific genetic characteristics, accelerating the development of improved crop varieties.
- **Plant genome sequencing and analysis**: The advent of high-throughput DNA sequencing technologies has revolutionized plant biotechnology by enabling the sequencing and analysis of entire plant genomes. This provides valuable insights into plant genetics, evolution, and gene function, aiding in the discovery of new genes and understanding complex biological processes.

Importance

Plant biotechnology has significant implications for agriculture, food production, and environmental sustainability. It offers the potential to develop crops that are more resistant to pests, diseases, and environmental stresses, leading to increased yields and reduced reliance on chemical inputs. Additionally, plant biotechnology plays a crucial role in the production of genetically modified organisms (GMOs), which have sparked debates around safety, ethics, and environmental impact.

Important Questions

1. What is plant biotechnology and its importance?
2. What is plant biotechnology and its applications?
3. Define the buffer solution.
4. What is pH value?
5. Describe to the genetic engineering.
6. What is MAB (Marker Assisted Breeding)

2 Carbohydrates

Carbohydrates are sugar molecules that also contain proteins and lipids. Which is one of the three primary nutrients present in foods and beverages. Our bodies convert carbohydrates into glucose. Glucose is the primary energy source for internal cells and organs, and it can be used immediately or stored in the liver and muscles for later use.

- Carbohydrates are generated by green plants during photosynthesis from carbon dioxide and water.
- The nucleic acids (DNA or RNA) are genetic information consisting of carbohydrates.
- Substances with the general formula $C_6H_{12}O_6$, such as wood, starch, and fiber contain carbon (C), hydrogen (H), and oxygen (O).
- The overall formula Many carbohydrates are frequently represented by $Cx(H_2O)y$.

Classification

There are mostly four primary groups.

- Monosaccharides
- Disaccharides
- Oligosaccharides
- Polysaccharides
- **Monosaccharide:** Monosaccharides are simple sugars present in a variety of fruits and honey, including grapes, bananas, and other fruits. The most common examples are five or six molecules linked together to form a chain-like molecule. Three of the most significant simple sugars, glucose (also known as dextrose, grape sugar, and maize sugar), fructose (fruit sugar), and galactose, have the same chemical formula ($C_6H_{12}O_6$) but differ structurally and in their properties; they are isomers.

```
 H H H H OHH   Aldehyde
 | | | | | |   Sugar
H-C-C-C-C-C-C-O
 | | | | |
 O O O O H
 H H H H
```

Glucose

```
 H H H OH   H
 | | | |    |
H-C-C-C-C-C-C-H
 | | | | |  |
 O O O H OOH
 H H H         Ketone
               Sugar
```

Glucose

- **Disaccharide**: its means combination of double sugar. A disaccharide is formed by two molecules of simple sugar joined together (Glucose + Galactose = Lactose). Sucrose, or table sugar, is a disaccharide composed of one molecule of glucose and one molecule of fructose found in sugar beets and sugarcane. Milk sugar (lactose present) and maltose are both disaccharides.
- **Oligosaccharide**: Oligosaccharides, which are made up of three to six monosaccharide units, are uncommon in nature, though a few plant counterparts have been identified.
- **Polysaccharide**: Polysaccharides are many different forms of sugar molecules that exist in nature as structural and energy-reserve carbohydrates. Polysaccharides are large molecules that can contain up to 10,000 monosaccharide units linked together. They vary greatly in size, structural complexity, and sugar concentration. Polysaccharides include cellulose, starch (in plants), and glycogen (in animals).

Anomers: it is an isomeric form of monosaccharide based on their configuration of hemiacetal or hemiketal.

Epimer or isomers or diastereomers: Two sugar molecules that differ in one position where the -OH group is located. D-glucose and D-galactose are the best examples of epimers, which differ only at the C-4 carbon. They are not diametrically opposed.

D-glucose and D-galactose are epimeric at carbon-4

```
   H\C=O                        H\C=O
  H-+-OH      Epimers          H-+-OH
 HO-+-H                       HO-+-H
  H-+-OH  <- Only one chiral -> HO-+-H
  H-+-OH     center with       H-+-OH
   CH2OH     different          CH2OH
             configuration
 D-Glucose                    D-Galactose
```

Enantiomers or Optical isomers: Enantiomers are two isomers that are related to one another through reflection. They are mirror images of each other that cannot be overlapped.

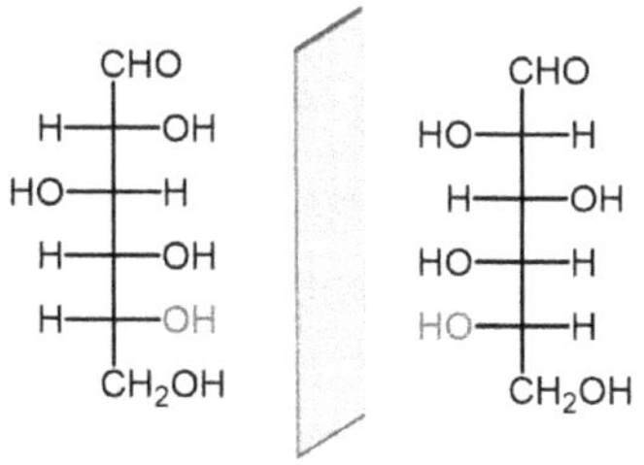

D-glucose L-glucose

Enantiomers

Diastereomers: Diastereomers are stereoisomers that have the same chemical formula. They are not exact copies of one another. They will have various physical and chemical qualities, such as different boiling/melting points, different hues, and may go through different chemical processes (typically, they go through the same chemical reactions but at different rates).

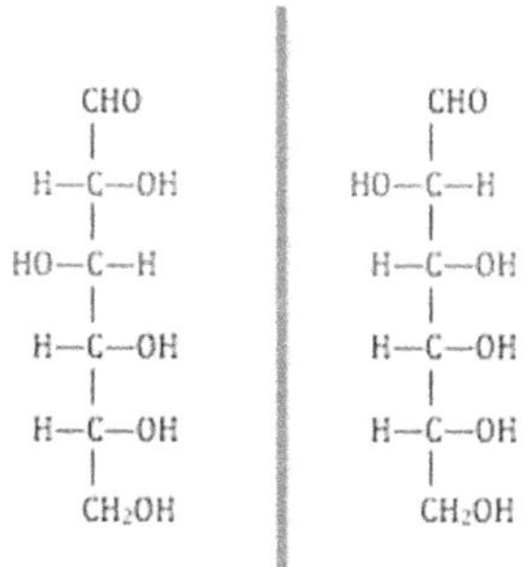

D-Glucose D-Altrose

Reducing and Non-reducing Sugar

S. No.	*Reducing Sugar*	*Non-Reducing Sugar*
1.	Reducing sugar has free ketone or aldehyde groups	Non-reducing sugar doesn't have any aldehyde or ketone group
2.	It exhibits mutarotation	It does not exhibit mutarotation
3.	Anomeric carbon has an OH group	Non-reducing sugars lack an OH group linked to the anomeric carbon.
4.	It can able to reduce solutions of various metallic ions	It cannot reduce the solution of various metallic ions
5.	Examples – Glucose, lactose, maltose etc.	Examples – Glycogen, sucrose, insulin etc.

Importance of Carbohydrates in Plants

- Carbohydrates are produced by plant during photosynthesis.
- In plants, starch sugar is stored as a carbohydrate.

- Photosynthesis occurs during the presence of sunlight.
- Plant's leaves are participated in the photosynthesis process.
- Glyceraldehyde 3-phosphate is produced during the Calvin cycle.
- Carbohydrates are converted into starch molecules which act as an effective energy store.
- Carbohydrates are converted into cellulose to build cell walls of plant.
- Carbohydrates are converted into Glucose which can be used for respiration to provide energy.
- Carbohydrates are converted to sucrose for transport in the phloem.
- Carbohydrates are converted into lipids as energy source in seed.
- Carbohydrates are converted into amino acid used to make protein.

Important Questions

1. What is reducing sugar?
2. What is anomers?
3. What is carbohydrates and its classification?
4. What is differnce between Epimer and Enantiomers?
5. Difference between Reducing and Non-Reducing sugar.
6. What is impotance of carbohydrates in plants?
7. Define the following terms:
   - i. Monosaccharide
   - ii. Disaccharide
   - iii. Oligosaccharide
   - iv. Polysaccharide

3 Lipids

Lipids are organic molecules that perform a different function in our body. cell membranes consist of two lipid bilayer structure and control to hydrophilic molecule across the membrane. Lipid help with moving and storing energy, absorbing vitamins and making hormones.

- **Hydrophilic means**- water-loving molecule
- **Hydrophobic means** - fat-solving molecule
- Unsaturated fatty acid has One or more double bond
- Saturated fatty acid has no double bond

Fatty acid: This is a simplest form of lipid consist from large number of lipid complex. Fatty acids are long chain of hydrocarbon (4 to 36 carbons long) and associate with carboxyl group. They are amphipathic nature means polar or non-polar as well as saturated and unsaturated fatty acids also.

a. Saturated Fatty Acid

- They have no double bond between C=C.
- Saturated fatty acid solid at room temperature.
- General formula of saturated fatty acid is CH_3-(CH_2)- COOH where n is the number of methylene groups between the methyl and carboxyl carbons.
- The example of Saturated fatty acid is Palmitic acid, Arachidic acid, Stearic acid, Myristic acid and Lauric acid.

b. Unsaturated Fatty Acid

- They have one or more hydrogen bond between C=C.
- They are of two type fatty acid such as monounsaturated and polyunsaturated fatty acid.
- Vegetable fat contains unsaturated fats like canola and corn oil.
- Monounsaturated fatty acid containing single bond i.e. olive and canola.

- In the case of polyunsaturated fatty acid, they have more than two double bonds i.e. soyabean oil, sunflower oil, corn oil and fish oil.
- Unsaturated fatty acid does not solid at room temperature. This appears in the form of liquid at room temperature because of the kinks that their double bonds induce in the molecules, fatty acid are frequently liquid at normal or room temperature.
- Examples of unsaturated fatty acid are Olic acid, Linoleic acid, Linolenic acid and Arachidonic acid.

Classification of lipids: There are three major groups of lipids

- **Triacylglycerols (Triglyceride):** They are consisting of three molecules of fatty acid and glycerol molecule. They are of two types such as simple triglyceride and complex triglyceride. They are non-polar (hydrophobic in nature) and major form of stored lipid.
- **Phospholipid**: This is an amphipathic nature constructed from fatty acid that is associated with the phosphate group. There are three major types: Phosphoglycerides, Glycerophospholipid and Sphingophospholids.
- **Glycolipids**: A glycolipid is a type of lipid molecule that contains a carbohydrate (sugar) chain attached to a lipid moiety. It is an important component of cell membranes and plays various roles in cell recognition, signaling, and cell-cell interactions. Glycolipids consist of a hydrophobic lipid tail and a hydrophilic carbohydrate head.

  Glycolipids are classified into two main categories based on their lipid component: glycosphingolipids and glyceroglycolipids.
- **Sterols**: Steroids are complex of triterpenes means composed of four rings are fused together. Cholesterol is an example of sterol present in animal not in plants and fungi. Stigmasterol, campesterol and sitosterol are an example of sterols present in plants and ergosterol present in fungi.

Membrane Lipid and Storage Lipid

Membrane lipids are a class of lipids that form the structural basis of cell membranes. They are amphipathic molecules, meaning they have both hydrophilic (water-loving) and hydrophobic (water-repelling) regions. Membrane lipids include phospholipids, glycolipids, and cholesterol.

- Phospholipids are the most abundant type of membrane lipid. They consist of a hydrophilic phosphate head group and two hydrophobic fatty acid tails. The arrangement of phospholipids in the cell membrane forms a lipid bilayer, with the hydrophobic tails facing inward and the hydrophilic heads facing outward, creating a barrier that separates the intracellular and extracellular environments.
- Glycolipids are another type of membrane lipid, characterized by a carbohydrate chain attached to a lipid moiety. They are mainly found in the

outer leaflet of the cell membrane and play a role in cell recognition, adhesion, and signaling.

- Cholesterol is a sterol lipid that is interspersed among phospholipids in the cell membrane. It helps regulate membrane fluidity and stability, and is involved in various cellular processes, including the formation of lipid rafts.

Storage Lipid

Storage lipids are a type of lipid molecule that serve as a long-term energy reserve in organisms. They are primarily composed of triglycerides, which consist of a glycerol molecule esterified with three fatty acid chains. Storage lipids are stored in specialized structures called adipocytes in animals and oil bodies in plants.

- Storage lipids provide a concentrated form of energy that can be utilized when needed. They are stored in an anhydrous (water-free) form, which allows for efficient energy storage without adding excess weight to the cell or organism. When energy is required, storage lipids are broken down through lipolysis, releasing fatty acids that can be oxidized to produce ATP.
- Unlike membrane lipids, storage lipids are not directly involved in the structure or function of cell membranes. Instead, they serve as an energy reserve, insulation, and protection for vital organs. They play a critical role in maintaining energy balance and providing fuel for metabolic processes.

Structures and properties of fatty acids: lipid molecule consist of hydrophilic head and hydrophobic tail. It presents in cell membrane as lipid bilayer form and occurs flip flop movement by them.

Properties

- Head portion of lipid molecule attached to the lipid bilayer in both side and hydrophobic tail occurs movement.
- Saturated and unsaturated fatty acid chain are attached to hydrophobic tail.
- Flip flop movement occurs in lipid bilayer and Hydrophobic molecule are passed from the cell membrane only.

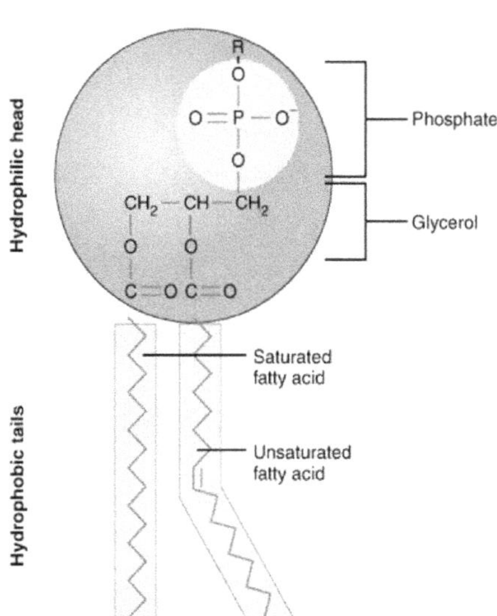

Important Questions

1. What is Lipid and its classification?
2. Define to the Triacylgycerds.
3. Difference between Saturated and Unsaturated fatty acid.
4. Describe the membrane and storage lipids with structure and properties.
5. What is difference between Hydrophilic and Hydrophobic molecule?
6. What is fatty acid? Discuss the difference between saturated and unsaturated fatty acids?
7. Write the short notes on:
   i. Gilycolipids
   ii. Lipoprotein
   iii. Steroids.

4 Vitamins

Vitamins are vital chemical components that play critical functions in the overall health and well-being of our bodies. They are required for a variety of physiological activities and are normally provided through a well-balanced diet.

What is a Well-balanced Diet?

A well-balanced diet that contains a range of fruits, vegetables, whole grains, lean meats, and healthy fats can supply the vitamins the body need. Dietary supplements may be helpful in some circumstances, but it is always better to consult your doctor before beginning any supplementing routine.

There are two main types of vitamins: water-soluble and fat-soluble.

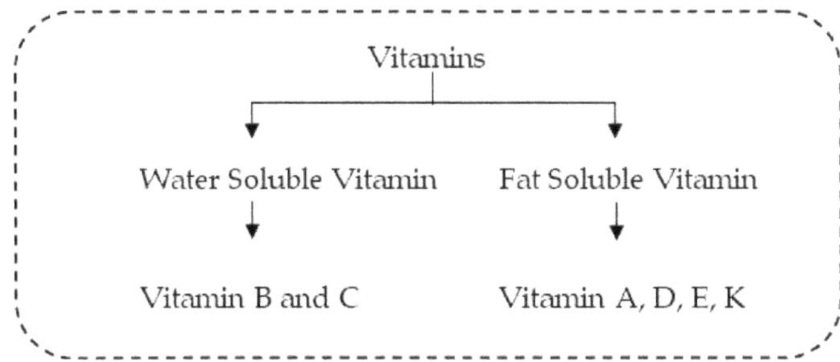

- **Water-Soluble Vitamins**: Vitamin dissolve in water and are not stored in the body. As a result, they must be supplied on a regular basis by food sources. These vitamins are necessary for energy production, immune system support, and skin and hair health.
- **Fate Soluble Vitamins**: Vitamins that dissolve in fat (for example, vitamins A, D, E, and K) can be stored in the body's fatty tissues and liver. Excess fat-soluble vitamins can build and potentially cause toxicity as a result of this storage capacity, thus it's critical to ingest them in adequate proportions.

Vitamins and their Specific Function

Each vitamin serves specific functions:

- **Vitamin A**: Vitamin A is necessary for good vision, immunological function, and healthy skin.
- **Vitamin B Complex**: B_1 (Thiamine), B_2 (Riboflavin), B_3 (Niacin), B_5 (Pantothenic acid), B_6 (Pyridoxine), B_7 (Biotin), B_9 (Folate), and B_{12} (Cobalamin) are all part of the vitamin B complex. These vitamins are essential for metabolism, neuronal function, and the synthesis of red blood cells.
- **Vitamin C**: Vitamin C is an antioxidant that boosts the immune system, promotes collagen formation, and improves iron absorption.
- **Vitamin D**: This Vitamin is essential for bone health because it aids in calcium and phosphorus absorption.
- **Vitamin E**: This vitamin is an antioxidant that preserves cells and boosts immunological function.
- **Vitamin K**: This vitamin is necessary for blood clotting and bone health.

Important Questions

1. What is vitamins?
2. What is water and fat soluble vitamins?
3. What is specific functions of different vitamins?

5 Amino Acid

Amino acid is compound which made from carbon, hydrogen, oxygen and nitrogen. It composed of an amino group and carboxyl group. In alpha amino acid, amino and carboxyl group are attached to same carbon atom which is called alpha carbon.

Amino acid act acid and base: Amino acid is amphoteric nature in which contain both group acid and base. Amino group is the basic group which act proton acceptor and carboxylic group is acidic group which act proton donor.

Isoelectric Point (pI): High proton concentration with low pH value and low concentration of proton with high pH at specific value called pI

Zwitterion: it is a compound of positively and negatively charged group. It doesn't have any electric charge.

Classification of amino acid: There are two major classifications

- Standard amino acid
- Non- standard amino acid

Standard amino acid: it based on polarity and chemical nature.

1. Based on Polarity:

a. Based on Non-polar in Nature–

- Glycine
- Alanine
- Leucine
- Tryptophane
- Isoleucine
- Phenylalanine
- Proline

- Methionine
- Valine

b. Based on Polar in Nature

Uncharged Polar Amino Acid

- Cysteine
- Serine
- Threonine
- Glutamine
- Asparagine
- Tyrosine

Charged Polar Amino Acid

Positive Charged Polar

- Histidine
- Arginine
- Lysine

Negative Charged Polar

- Aspartate
- Glutamate

2. Based on Chemical Nature of R Group

- **Aliphatic amino acid**: glycine, alanine, valine and leucine
- **Aromatic amino acid**: Phenylalanine, tyrosine, and tryptophane
- **Hydroxyl amino acid**: Serine and threonine
- **Carboxylic amino acid**: Aspartate and Glutamate
- **Sulphur containing**: Cysteine, cystine and methionine
- **Amino acid group**: Lysine and arginine
- **Imino acid group**: Proline
- **Amide amino acid group**: Asparagine and Glutamine

Non Standard amino acid: Some of them amino acid that does not ribomically incorporated into a protein called non standard.

4-Hydroxyproline, 5- Hydroxyproline, N-acetylserine, N- formylmethionine

Selenocysteine amino acid: 21st standard amino acid and Code by UGA (stop codon)

Pyrrolysine amino acid: 22nd standard amino acid and Code by UAG (stop codon)

Titration

- Effect of pH on amino acid called as titration of amino acid.
- Titration means considered two group first is α amino and second is α carboxylic group.
- It depends on protonated and deprotonated form.
- In strong base, alanine loses two protons stepwise. net charged -1, carboxyl group is deprotonated.
- In strong acid, carboxyl group are uncharged in stepwise. net charged +1, amino group is protonated.

Isoelectric Point (pI)

Titration of amino acids is a chemical process used to determine their acid-base properties. Amino acids contain both acidic carboxyl (-COOH) and basic amino (-NH_2) functional groups. The titration curve of an amino acid illustrates the change in pH as a strong base, usually sodium hydroxide (NaOH), is gradually added to a solution containing the amino acid.

The titration process involves the following key points:

Initial pH: At the beginning of the titration, the amino acid solution has an acidic pH due to the presence of the carboxyl group, which can release a hydrogen ion (H+). The amino group, however, has a basic effect that slightly elevates the pH.

First equivalence point: As the strong base is added, it reacts with the carboxyl group, causing it to lose its acidic properties. This occurs at the first equivalence point, known as the pKa1, where the amino acid acts as a monoprotic acid, losing a single proton. At this point, the pH increases sharply.

Buffer region: After the first equivalence point, the solution enters a buffer region, where the pH rises more gradually. In this region, the amino acid functions as a zwitterion, with both the carboxyl and amino groups neutralized.

Second equivalence point: As more base is added, the amino group starts to deprotonate, resulting in the second equivalence point (pKa2). The pH rises sharply again as the solution becomes more basic.

Isoelectric point (pI): The isoelectric point is the pH at which the amino acid is electrically neutral. It occurs between the two equivalence points when the acidic and basic groups are fully deprotonated. At the isoelectric point, the amino acid has no net charge.

The titration curve of an amino acid provides valuable information about its acid-base properties, including the pKa values of the carboxyl and amino groups. These values can vary among different amino acids, influencing their behavior and interactions in biological systems.

Overall, the titration of amino acids allows for the determination of their acid-base behavior, providing insights into their structure, properties, and reactivity.

$$pI = \frac{pK1 + pK2}{2}$$

pK value: The pKa value is a measure of the acidity or basicity of a compound, particularly acids and bases. It quantifies the degree of ionization of a compound in a solution and indicates the pH at which half of the compound is in its protonated (acidic) form and half is in its deprotonated (basic) form.

The pKa value is based on the equilibrium constant of the dissociation reaction of an acid or base in water. It represents the negative logarithm (base 10) of the equilibrium constant (Ka) for the reaction:

$$pKa = -\log10(Ka)$$

Important Questions

1. What is Amino acid and its classification based on Polar and Non Polar?
2. What is chemicals nature of amino acid?
3. Describe the Standard and Non-Standard amino acid.
4. What is Titration of amino acid ?
5. What is Isoelectric point (pI)?
6. What is pk value?
7. Define the following :
   i. Sulphar containing Amino acid
   ii. A positivily charged and negative changed amino acid.
   iii. A purine and a pyrimidine base with their chemical names.

6 Proteins

A peptide bond is a chemical connection formed between the carboxyl group (COOH) of one amino acid and the amino group (NH_2) of another. It is an important link in the creation of proteins. A peptide bond is formed through a condensation reaction, also known as a dehydration reaction. A molecule of water is removed during this process as the carboxyl group of one amino acid interacts with the amino group of another amino acid. This reaction is catalyzed by enzymes in ribosomes of living organism. The peptide bond joins two or more amino acids together, releasing a water molecule from each peptide bond.

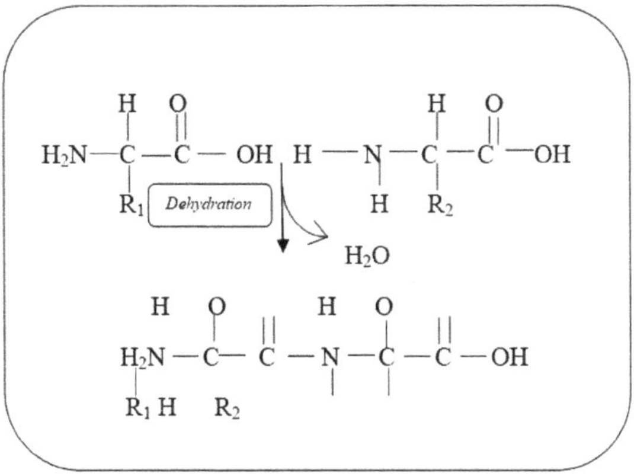

Formation of peptide bond

Ramachandran plot: The Ramachandran plot is a graphical representation of the torsional angles [φ (phi) and ψ (psi)] of amino acid residues in a protein structure. It was developed by G.N. Ramachandran and his colleagues in the 1960s. The plot allows us to analyze and visualize the conformational angles of the protein backbone, which play a crucial role in determining the protein's overall structure and stability.

The x-axis of the Ramachandran plot represents the φ (phi) angle, which is the rotation around the α-carbon to the peptide bond. The y-axis represents the ψ (psi) angle, the rotation around the α-carbon to the next peptide bond. The plot is typically divided into regions that correspond to different secondary structure elements such as alpha helices, beta sheets, and coil regions.

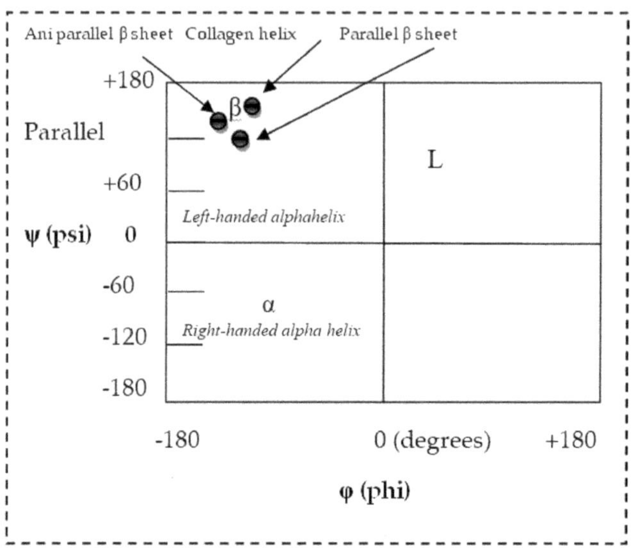

By plotting the φ and ψ angles of each amino acid residue in a protein, researchers can identify favorable and unfavorable regions of conformational space. The favorable regions indicate allowed conformations, while the unfavorable regions indicate steric clashes or strained conformations that are less likely to occur. The Ramachandran plot is an indispensable tool for protein structure determination, refinement, and validation. It has greatly contributed to our understanding of protein folding, stability, and functional properties. Researchers use this plot to assess the quality of protein structures, identify structural anomalies, and refine computational models.

Rotation
(Phi) φ
(Psi) Ψ
H
Cα
R1
O
C
N
H
H
Cα
R2
C
O
H
N
H
Cα
R3

Amino Acid
Amino Acid
Amino Acid

Rotation of peptide bond: The rotation is permitted between N-Cα and Cα-C bond. Where N is amino group of fist amino acid, Cα is alpha carbon present in central of amino acid and C is part of carboxyl group of another amino acid. This angle is labeled from phi and psi. this rotates between -180 to +180. Ramachandran plot is usually indicated to 2D map of protein.

Properties: There are major properties of protein

- Each protein has two end N terminal and C terminal.
- N terminal has amino group and C for carboxylic group.
- Protein is amphoteric in nature.
- Glycine is neutral amino acid because he does not carry to the charge.
- Largest and smallest subunit of ribosome read to the mRNA strand with the help of tRNA carry to nucleotide and arrangement for making of protein.
- AUG or GUG are initiation codon.
- UAA, UAG and UGA are stop codon.
- UGA codon code to selenocysteine.
- UAG codon code to Pyrrolysine.
- If pH high pI will be low.
- If pH low, pI will be high.
- If pH is equal to pI value called as pKa value means no charge (+ or -) carries.

On the Basis of Protein Structure:

- **Primary Structure**: Amino acid are arranged in the sequence form that is primary structure.
- **Secondary Structure**: the most common type of secondary structure are alpha helix and beta pated sheet.

  α-Helix: it is a rigid and road like structure that form when polypeptide chain twist into conformation.

  β-pleated sheets: It is a anti parallel form of two or more polypeptide chain segment and form a row of hydrogen bond to each other.
- **Tertiary structure**: it refers to the three-dimensional conformations of the polypeptide chain. There are major interaction between polypeptide chain such as hydrophobic interactions, electrostatic interaction, hydrogen bond, van der Waals interaction and covalent interaction.

  Covalent interaction: it presents in tertiary structure is interchain disulfide bond between two molecule of cysteine amino acid residue due oxidation of sulfhydryl group.
- **Quaternary Structure**: it is type of multimeric protein structure which Composed of two or more polypeptide chains such as Hemoglobin and others

On the Basis of Composition:

- **Simple Protein:** Simple proteins have only amino acids as their chemical building blocks and no other chemical groups. They are formed through the linkage of amino acids by peptide bonds. Examples of simple proteins include albumin, globulins, and histones.
- **Conjugated Protein:** Conjugated proteins, also known as complex proteins, are composed of amino acids as well as non-protein components called prosthetic groups. These prosthetic groups can be organic molecules like carbohydrates or lipids, or inorganic molecules like metal ions. Conjugated proteins exhibit specialized functions due to the presence of these additional components. Examples include hemoglobin (contains heme prosthetic group) and lipoproteins (contain lipids).
- **Fibrous Protein:** Fibrous proteins have a long, elongated shape and are typically insoluble in water. They provide structural support and are commonly found in connective tissues, tendons, and muscles. Fibrous proteins are characterized by a repetitive pattern of amino acid sequences, which allows them to form strong fibers. Examples include collagen (found in skin, bones, and cartilage) and keratin (found in hair, nails, and skin).
- **Globular Protein:** Globular proteins have a compact, spherical shape and are soluble in water. They play diverse roles in biological processes such as enzyme catalysis, transport of molecules, and immune responses. Globular proteins typically have complex three-dimensional structures, and their amino acid sequences are folded into specific arrangements. Examples include enzymes like amylase, antibodies, and hemoglobin (when not considering its prosthetic group)

Important Questions

1. Define the Peptide bond.
2. What is Ramachandran plot.
3. How does the Rotation of peptide bond.
4. Describe the classification of protein structure.
5. What is Globular protein?

7 Enzyme

Enzyme is a biomolecule substance which act as catalytic process in living organism. All the cellular process are mediated by enzymes. Enzyme have several properties

- The biological processes that occur within all living organisms are chemical reactions regulated by enzymes.
- It catalyzes to the metabolism process which are includes such as glycolysis, pyruvate oxidation citric acid cycle and electron transport chain for the one molecule glucose oxidation to ATP yield.
- Large nutrient molecules such as proteins, carbohydrates, and fats are breakdown into smaller molecules.
- All enzymes are protein but all protein are not enzymes.
- It does not change equilibrium state of biochemical reaction, it changed to equilibrium rate of chemical reaction.
- It increases the rate of reaction by lowering activation energy and it are highly specific biomolecule.

Classification of Enzymes:

There are Two Major Groups

- **Simple Enzymes**: it consists of amino acid arrangement formed protein
- **Conjugated Enzyme**: it consists of protein and as well as non protein are formed conjugated protein.

Conjugated = Protein component + Non Protein component (cofactor such as Ca^{2+}, others)

Cofactor: The metal ion has catalytic activity which contain non protein components such as Ca^{2+}, Mg^{2+}, Zn^{2+}, Mn^{2+}, K^{+}, Ni^{2+}, Cu^{2+}, etc.

Coenzymes: Cofactors are complex of organic compounds they can play the various role for enhancement in catalytic activity such as vitamins and others.

Prosthetic group: Protein are tightly linked with coenzyme (such as vitamin) by covalently or non-covalently called prosthetic group.

Prosthetic groups= Protein + Coenzyme (such as vitamin)

Apoenzyme: Protein component is biological inactive. it doesn't link with cofactor called apoenzyme

Holoenzyme: Protein component is biological active because it linked with cofactor called holoenzyme.

Simple enzyme + cofactor (such as Ca^{++}, Mg^{++} etc.)

Enzyme Naming: Most of enzyme named for their substrates and catalyzes reaction

ATPase: Help in breakdown ATP

ATP synthase: Help in synthesis of ATP

EC Number: EC stand for Enzyme commission number such as EC 2.7.1.2 (hexokinase)

First three number (2.7.1) denote to the major class, subclass, and sub-subclass and last number (2) denote to the serial number of sub-subclasses.

Classification of Enzyme

There are Six Major Classes

- **Oxidoreductase**: Catalyze to oxidation – reduction process such as peroxidase, oxidase, etc.
- **Transferases**: Group transfer from one to another molecule such as kinase, phosphorylase etc.
- **Hydrolases**: Catalyze to bond such phosphodiesterase, amide bond and release water molecule such as peptidase, phosphatase, etc.
- **Lyases**: Catalyze to C-O, C-C, C-N, C-S and other bond. It involves elimination process such as synthase, aldoses etc.
- **Isomerases**: Catalyze to intermolecular rearrangements yield isomeric form such as mutases
- **Ligases**: catalyze to the formation of bond by hydrolysis of ATP such as carboxylases

How to Operate Enzyme

There are two major model

- **Lock and Key Model:** in this model, shape of substrate and binding of enzyme is lock and Key based. It is very specificity reaction between enzyme and substrate.

- **Induced fit Model:** in this model, the enzymes are flexible in nature and shape change during the binding with substrate molecule. The binding of substrate induced a conformational change in enzyme called induced fit model.

Michalis-Menten Model

This model is associated between reaction velocity (rate of enzyme reaction) and substrate concentration. The model was first discovered by Leonor Michaelis and Maud Menten in 1913. They studied the reaction of invertase enzyme which hydrolyzed the sucrose into glucose and fructose

How to work model: In this model, different substrate concentration is used and measure reaction velocity. Rate of reaction depend on the different concentration of substrate [S].

There are two major activities:

- If the substrate concentration low, initial velocity increase
- If the substrate concentration increase, reaction of maximum velocity

The substrate concentration at which reaction of rate reaches high its maximum velocity called as k_m (Michalis constant).

Features: There are two major significant features

- In the case of substrate concentration was kept constant and enzyme concentration was increased, reaction rate increased.
- In the case of enzyme concentration was kept constant and substrate concentration was increased, reaction rate increased in maximum value.

Enzyme Reaction Formula

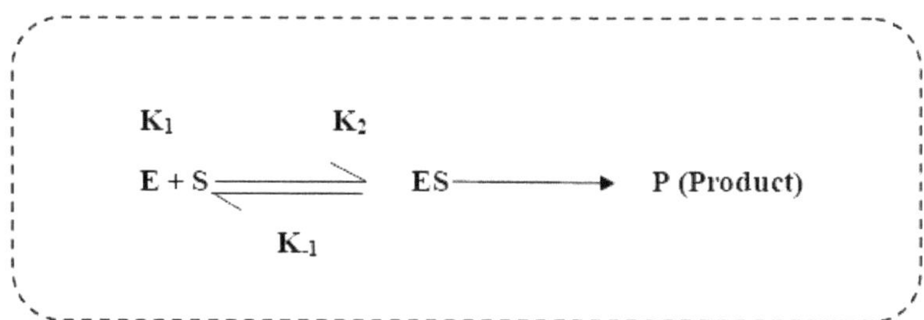

E= Enzyme, S= Substrate, ES= Enzyme-substrate complex, P= product

K_1=reaction rate, k_{-1}= reverse rate of reaction, K2= converse from ES to P(product)

Important Questions

1. What is Enzyme and its classification with properties?
2. What is Cofactor?

3. What is Coenzyme ?
4. What is Apoenzyme and Holoenzyme?
5. What is Michalis- Menten Model?

8 Enzyme Inhibition

There are three reversible Enzyme inhibitor

1. Competitive Inhibition

Km increase

V_{max} unaffected

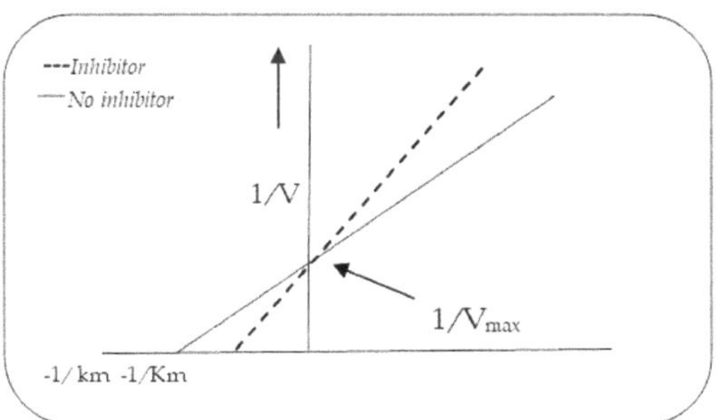

2. Non-competitive Inhibition

Km unaffected

V_{max} reduced

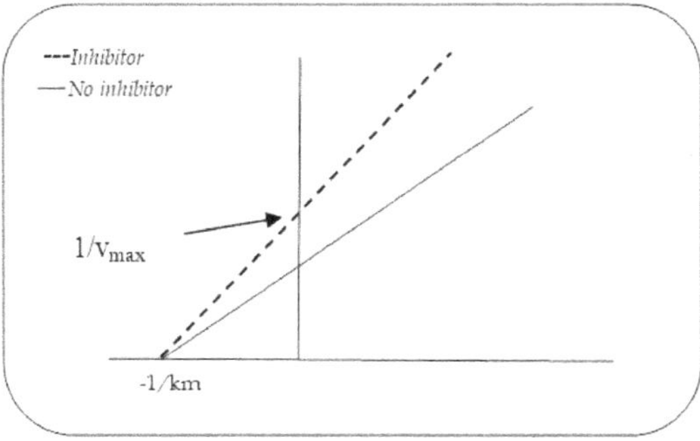

3. Uncompetitive Inhibition

KM reduced

V_{max} reduced

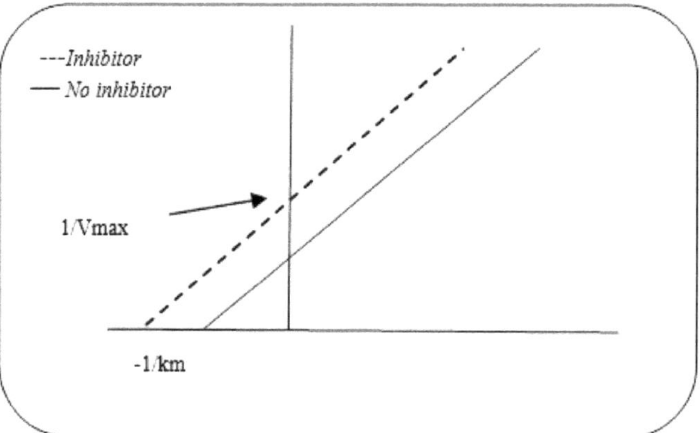

Important Question

1. What are difference between Competitive, Non competitive and Uncompetitive inhibition?

9 Nucleic Acid

Nucleic acids are chemical substances that are essential for protein synthesis and act as the primary information-carrying molecules in cells. Nucleic acids are classified into two types: deoxyribonucleic acid (DNA) and ribonucleic acid (RNA).

Nucleotide: The chemical composition of nucleotide is consisting of a phosphate group, a sugar and a nitrogenous base.

Nucleoside: The chemical composition of nucleoside are consisting of a sugar and a base without the phosphate group.

Difference between DNA or RNA Chemical Structure

- DNA means: deoxy means oxygen absents at 2 carbons
- RNA means: Oxygen present at 2 carbons

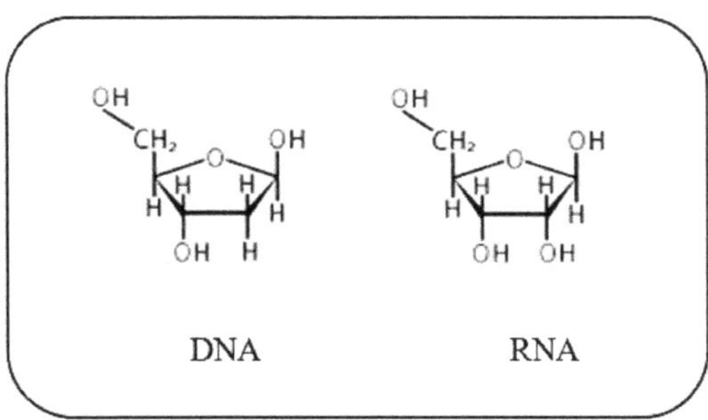

Difference between DNA or RNA

S. No.	*DNA*	*RNA*
	DNA stand for deoxynucleic acid	Ribonucleic acid
	It consists of three components like- phosphate + sugar + base	It consists of two components like – sugar + base
	Double standard	Singal standard
	It contains deoxyribose sugar molecule.	It contains ribose sugar molecule.

S. No.	*DNA*	*RNA*
	The nucleotide are adenine, guanine, cytosine and thymine. But uracil is not present	Uracil is present instead by thymine.
	It is stable and not hydrolyzed easily by alkalis.	It is unstable and hydrolyzed easily by alkalis.

Nucleotide and their structure: There are consist of Phosphate group + Sugar + Nucleotide base.

Phosphate group (PO_4): A phosphate group is backbone of the DNA double helix that provides structural support.

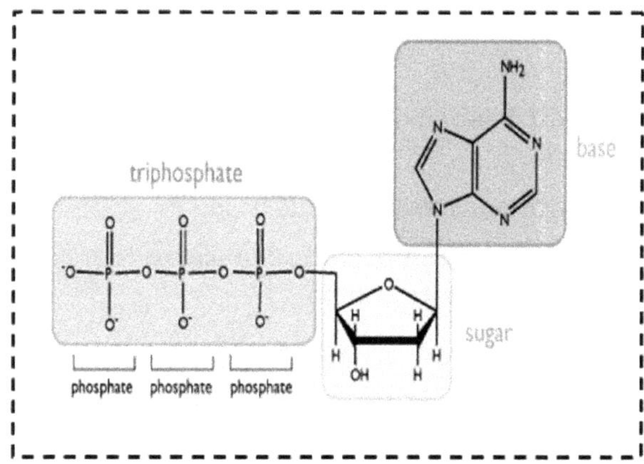

Sugar: DNA have deoxyribose sugar molecule in which one oxygen molecule are absent at second carbon number of sugar molecule. Sugar molecule are attached to both side phosphate and base.

Pyrimidines

Cytosine
C

Thymine
T

Purines

Guanine
G

Adenine
A

Nucleotide base: There are four nucleotide which involve for making DNA double helix structure. Adenine (A), Guanine (G), Cytosine (C) and Thymine (T) are made to DNA structure.

Difference between A, B and Z DNA

S. No.	Characteristics	A-DNA	B-DNA	Z-DNA
	Helix	Right-handed helix	Right-handed helix	Left-handed helix
	Diameter	Diameter of 26Å	20Å	18Å
	base pair per helical turn	11 base pair	10 base pair	12 base pair
	Helix rises per base pair	2.3 Å	3.3 Å	3.8 Å
	Major Grove	Wild deep	Narrow deep	Flat (absent)
	Minor Grove	Narrow, Shallow	Broad, Shallow	Narrow, deep
	Sugar pucker	C3′-endo	C2′-endo	C2′-endo and C3′-endo
	Conformation of glycosidic bond	Anti	Anti	Anti for pyrimidines and Syn for purine

Chargaff rule: it defines to the DNA of any living organism which have 1:1 ratio in the pairing of nucleotide form like purine and pyrimidine. In this rule, guanine is equal to cytosine and adenine is equal to thymine

Note: Chargaff rule is not applicable in single stranded DNA or RNA

Adenine (A) = Thymine (T)

Guanine (G) = Cytosine (C)

Purine (A G) = Pyrimidine (T, C)

DNA structure: DNA double helical structure was proposed by J.D Watson and F.H.C. Crick (1953).

There are three major structures of DNA

- **Primary Structure**: The arrangement of nucleotide base is in single-stranded form
- **Secondary Structure**: it is double-stranded helical structure like- A DNA, B DNA and Z DNA.
- **Tertiary Structure**: DNA structure present is in supercoiled form

Important Questions:

1. What is Nucleic Acid and describe the Nucleotide and Nucleoside?
2. Difference between DNA and RNA.
3. Difference between A-DNA, B-DNA and Z-DNA.
4. Describe the Chargaff rule and DNA structure also

10 Carbohydrates Metabolism

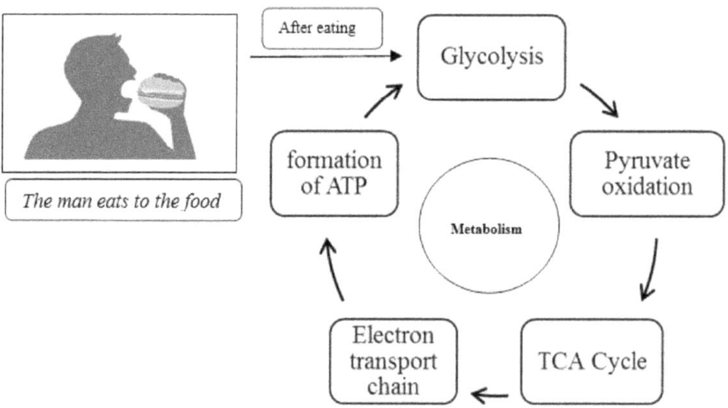

Metabolism: Chemical reaction occurs in the living cells called as biochemical factories. All the interconnected biomolecule reaction occurring within cell are catalyzed by enzymatic action called as Metabolism. Each step of metabolic pathway is catalyzed by specific enzyme.

Starting Molecule → Product

A —*Enzyme 1* (Reaction 1)→ B —*Enzyme 2* (Reaction 2)→ C —*Enzyme 3* (Reaction 3)→ D

Metabolic pathway based on two categories

- **Anabolic pathway (endergonic process):** it means synthesis of compounds and consume energy. Synthesis of amino acid and polypeptide are example of endergonic process.
- **Catabolic pathway (exergonic process):** it means oxidative breakdown of large molecule and release energy called as exergonic process. Glucose breakdown is the example of exergonic process.

$$\text{Glucose} \xrightarrow{\text{Breakdown}} \text{lactic acid (skeletal muscle)}$$

- **Amphibolic pathway:** Both process anabolic and catabolic pathway are work in this pathway. TCA is example of amphibolic pathway

Metabolic pathway is work on the three forms

- Linear (Such as Glycolysis)
- Cyclic (Such as TCA or Citric Acid Cycle)
- Spiral (Such as fatty acid biosynthesis)

Respiration: it is a metabolic process in which free energy released from the oxidation of organic compounds. The organic compound is used in the formation of ATP during the process of respiration are known as respiratory substance for example - carbohydrates, fats, protein or organic acids etc. Respiration are two types: aerobic respiration, anaerobic respiration.

Aerobic Respiration in Metabolism

- During this process, when molecule of glucose (180 g) is completely oxidized into H_2O and CO_2 with 686 kcal (kilocalories) called respiration process. It will further used for synthesis of ATP with the help of glycolysis, pyruvate oxidation, TCA cycle and electron transport chain work step by step form in our body.
- $C_6H_{12}O_6 \longrightarrow 6\,CO_2 + 6H_2O$ + Energy (ATP + Heat)
- It is based on oxygen depend process.
- It involves the electron transport system.
- ATP synthesis occurs through substrate level phosphorylation and oxidative phosphorylation.

Anaerobic Respiration in Metabolism

- It is based on oxygen independent process. Anaerobic compound such as sulphate, nitrate, Sulphur etc.
- It involves the electron transport system.
- ATP synthesis occurs through substrate level phosphorylation and oxidative phosphorylation.

Fermentation Process in Metabolism

- It is a type of aerobic respiration process in living cells like some of bacterial cells and others.
- It does not involve electron transport system.
- ATP synthesis occurs through substrate level phosphorylation only.

Cellular Component of Metabolic Pathway

S. No.	Process	Cellular components
	A. In eukaryotes	
	Glycolysis	Cytosol
	Citric Acid Cycle or TCA cycle	Mitochondria
	Pyruvate oxidation	Mitochondrial Matrix
	Oxidative phosphorylation	Inner mitochondria membrane
	B. In Prokaryotes	
	Glycolysis	Cytosol
	TCA	Cytosol
	Oxidative phosphorylation	Plasma membrane
	Fermentation	Cytoplasm
	Cori cycle (Lactic acid cycle)	Recycled to glucose in the liver
	Glyoxylate cycle (Convert Acetyl - CoA to Succinate)	Glyoxysome in plants but absent in animal

1. Glycolysis

It means lysis of sugar molecule

It is also known as **Embden – Meyerhof pathway**

It is a process in which one molecule of glucose is partially oxidized into the two molecules of pyruvate through the enzymatic action.

Steps of glycolysis

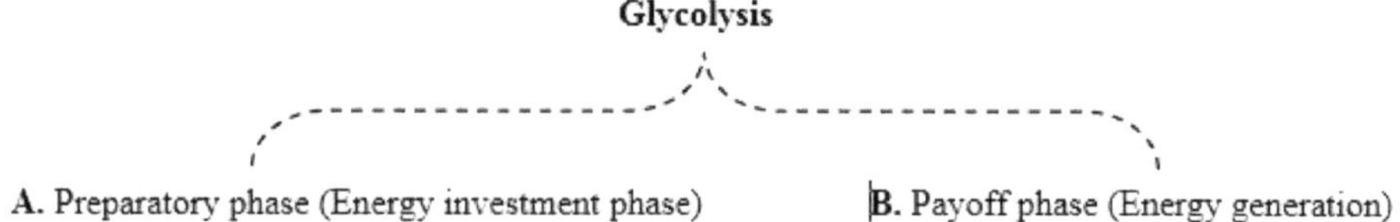

A. Preparatory Phase (Energy Investment Phase)

Step 1 (Phosphorylation): Glucose is phosphorylated by ATP to form of Glucose 6 phosphate. It process is catalyzed by hexokinase enzyme. It is a irreversible process

Step 2 (Isomerization): it is a reversible process. In this process, glucose 6 phosphate is isomerized into fructose 6 phosphate though the phosphoglucoisomerase enzyme.

Step 3 (Phosphorylation): Fructose 6 phosphate is phosphorylated by ATP to fructose 1,6- bisphosphate. It process is catalyzed by phosphofructokinase-1(PFK-1). This process is irreversible process.

Step 4 (Cleavage): The fructose 1,6- bisphosphate is cleaved and produced into two molecule i.e. Glyceraldehyde-3-phosphate (3C) and Dihydroxyacetone phosphate (3C). It process is catalyzed by Aldose and Zn^{++}.

Step 5 (Isomerization): Dihydroxyacetone phosphate (3C) further isomerized into Glyceraldehyde-3-phosphate (3C) through the triose phosphate isomerase. Formation of 2 molecule of Glyceraldehyde-3-phosphate (3C) in preparatory phase.

B. Payoff Phase (Energy Generation)

Step 6: Two molecule of Glyceraldehyde-3-phosphate (3C) are oxidized into 2 × 1,3- bisphosphoglycerate (3C) catalyzed by Glyceraldehyde-3-phosphate dehydrogenase.

Step 7: Formation of ATP by substrate level phosphorylation. In this step 2 × 1,3- bisphosphoglycerate (3C) catalyzed by phosphoglycerate kinase produced 2 × 3 phosphoglycerate (3C).

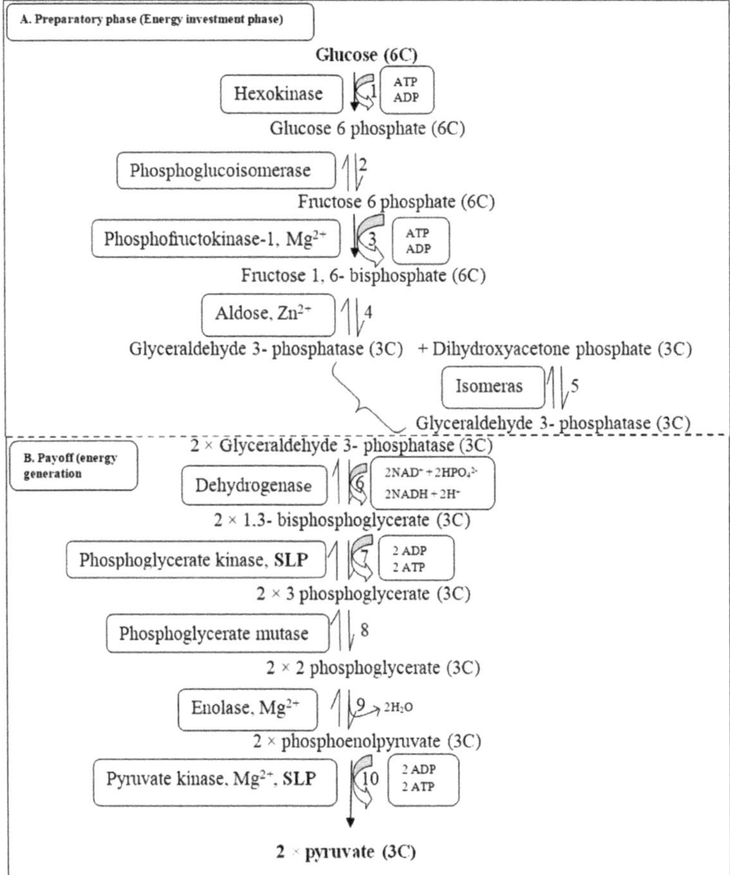

***SLP:** Substrate Level Phosphorylation, **6C:** Six carbon molecule, **3C:** Three carbon molecule*

Step 8: 2 × 3 phosphoglycerate (3C) are catalyzed by phosphoglycerate mutase enzyme produced 2 × 2 phosphoglycerate (3C).

Step 9: 2 × 2 phosphoglycerate (3C) is catalyzed by enolase and release $2H_2O$ produced 2 × phosphoenolpyruvate (3C). this process is inhibited by fluoride

Step 10: The last step of glycolysis is a substrate level phosphorylation. in this step, transfer phosphoryl group from phosphoenolpyruvate to ADP is catalyzed by pyruvate kinase produced two molecules of pyruvate. This process is a irreversible process.

ATP Synthesis by Glycolysis

Glucose + $2NAD^+$ + 2 ADP + $2HPO_4^{2-}$ ⟶ 2 Pyruvate + 2 NADH + 2 ATP + 2 H_2O

Total Energy yield

In the presence of oxygen:

- 2 ATP + 2 NADH, 1 NADH carries to 2.5 ATP
- TOTAL ATP: 2 ATP + 5 ATP (NADH) = 7 ATP

In the absence of oxygen

- 2 ATP by substrate level phosphorylation
- NADH absent

2. Pyruvate oxidation: Oxidation of Pyruvate molecule occurs in mitochondria matrix through pyruvate dehydrogenase produced acetyl-CoA. This process occurs in the presence of oxygen.

Pyruvate dehydrogenase has the multicomplex enzyme play the major role in oxidative decarboxylation process. There are three major enzymes

- Pyruvate dehydrogenase (E_1)
- Dihydrolipoyl transacetylase (E_2)
- Dihydrolipoyl dehydrogenase (E_3)

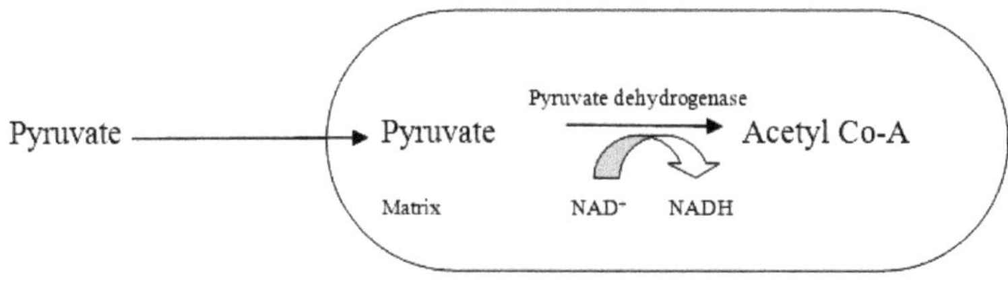

Mitochondria

Total energy: 2 NADH (2 × 2.5 = 5ATP)

Citric Acid Cycle or TCA: It is also called Krebs cycle or tricarboxylic acid cycle. In this cycle, one molecule of Acetyl CoA (3C) entering into citric acid cycle produced $2CO_2$ + 1 ATP + 3 NADH + 1 $FADH_2$. TCA was discovered by H. A. Krebs (British Biochemist) who received the novel prize in 1953. This cycle occurs in mitochondria matrix of eukaryotic cell.

Total energy:

- $4CO_2$ + 2 ATP + 6 NADH + 2 $NADH_2$
- 1 NADH = 2.5 ATP, 1 $FADH_2$ = 1.5 ATP
- 2 ATP + 6 NADH (6 ×2.5=15 ATP) + 2 $NADH_2$ (2×1.5 = 3 ATP) = 20 ATP

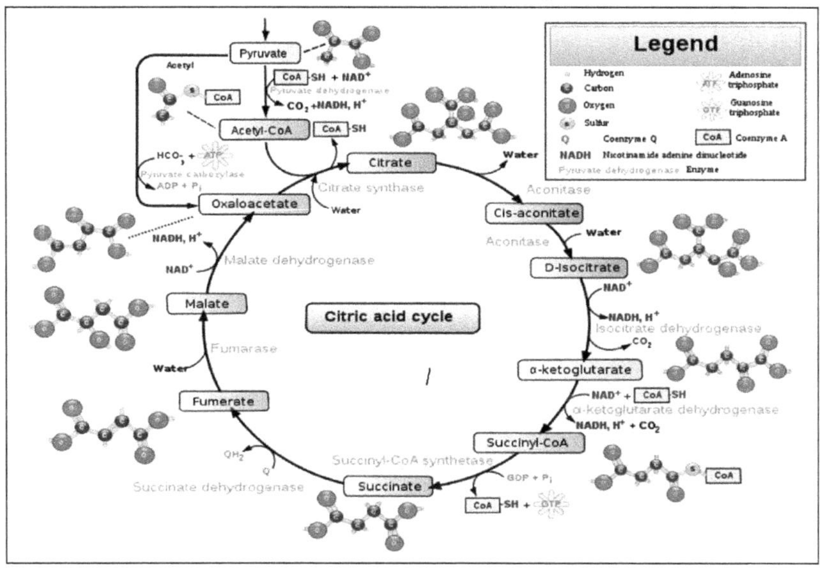

Fig. 1: Steps of TCA Cycle

Link: *By Narayanese, WikiUserPedia, YassineMrabet, TotoBaggins/CC BY-SA 3.0/Wikimedia Commons*

Steps of TCA: The major steps are involves in TCA cycle

Step 1: Condensation: In this step, Acetyl-CoA and water to yield citrate and coenzyme A. This is catalyzed by *citrate synthase.*

Step 2a: Dehydrogenase: Remove of water in this step and this reaction is formed cis-aconitate catalyzed by aconites . Aconitase is inhibit by fluoroacetate (it is a type of pesticide and this reaction block by them).

Step 2b: Hydration: Addition of water in this step and catalyzed by *aconitase*

Step 3: **Oxidative decarboxylation**: This step is catalyzed by isocitrate dehydrogenase and formed α-ketoglutarate from the oxidation of isocitrate.

Step 4: **Oxidative decarboxylation**: The formation of succinyl-CoA from α-ketoglutarate. This step is catalyzed by α-ketoglutarate dehydrogenase and produced NADH , CO_2.

Step 5: **Phosphorylation**: succinyl-CoA break into succinate and produced GTP or ATP. This is catalyzed by succinyl-CoA synthetase. This step is also called substrate level phosphorylation.

Step 6: **Dehydrogenation**: Formation of fumarate from removal of two hydrogen atom of FAD convert into FADH2 and this is catalyzed by succinate dehydrogenase. Malonate is the inhibitor of succinate dehydrogenase.

Step 7: **Hydration**: in this step, addition of water to hydroxyl group of fumarate and convert into malate. This is catalyzed by fumarase.

Step 8: **Dehydrogenation**: This is the last step of TCA cycle. In this step, malate convert into oxaloacetate and NAD+ into NADH. This is catalyzed by malate dehydrogenase. And again, start this cycle from oxaloacetate to next considered as first step. The regenerating of oxaloacetate is needed for the starting TCA cycle from first step.

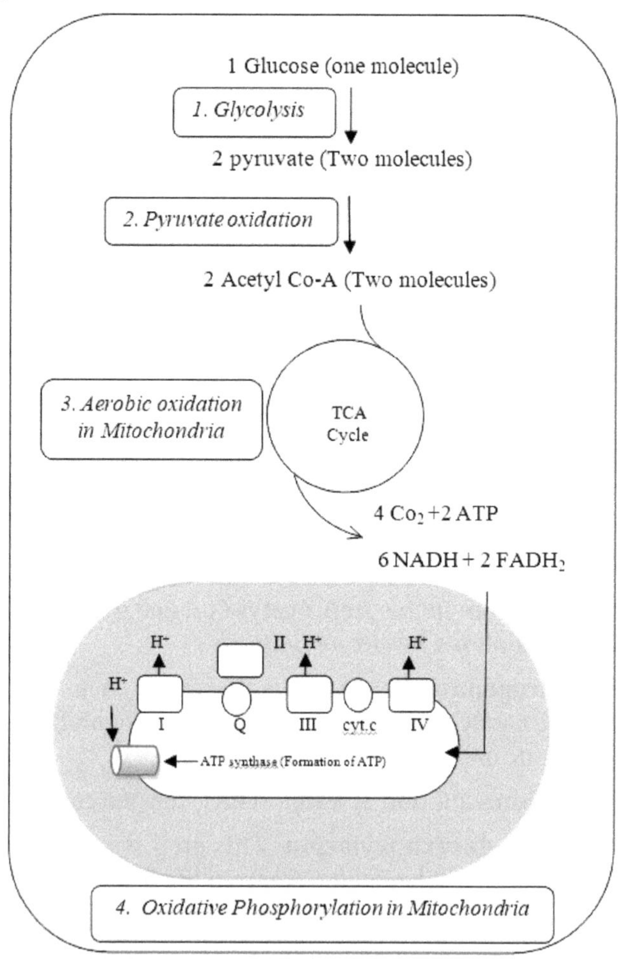

Fig. 2: Overall reaction: *One molecule of glucose is oxidized and produced total 32 ATP*

Oxidative Phosphorylation (Electron Transport Chain): Free energy are released from oxidation of glucose in the form of CO_2, NADH and FADH during glycolysis, pyruvate oxidation and citric acid cycle. Electron are released from NADH and $FADH_2$ to O_2 molecule through the unidirectional electron transport chain present in inner mitochondria membrane space. Electron are transformed to O_2 forming two molecules of H_2O.

There are four major respiratory complexes

- NADH coenzyme Q reductase (Complex I)
- Q cycle (carrier protein)56
- Succinate coenzyme Q- reductase (Complex II)
- Coenzyme Q cytochrome c reductase or Cytochrome bc1 complex (complex III)
- Cytochrome – c (carrier protein)

4. Cytochrome C Oxidation (Complex IV)

Electron Transport through the Respiratory Complex:

- Hydrogen ion are removed from NADH/ $FADH_2$ converted into oxidized form NAD^+/FAD^+ released one proton and 2e-electron but $FADH_2$ does not release protons during transport of electron across the inner mitochondria membrane.
- One proton is transported from mitochondrial matrix to inner mitochondrial membrane space. Total four electron are sequentially transferred to O_2 molecule formed two H_2O molecule.

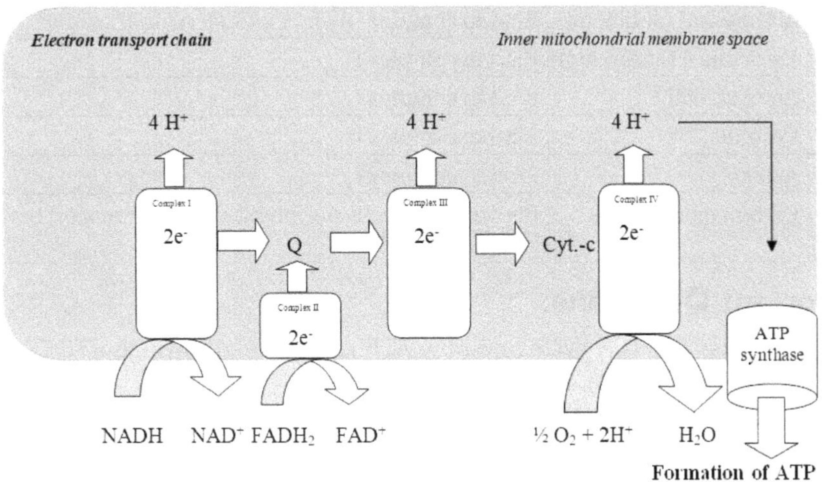

Fig. 3: Electron transport chain in inner mitochondria membrane space

ATP Synthase:

Proton motive force is used for formation of ATP by ATP synthase. ATP synthase are multimeric protein complex present in inner mitochondria membrane. It consists of two components F_1 and F_0. F_1 component consist one a subunit two b subunit and 9 to 12 subunit of c. F_0 is a transmembrane complex that regulated H^+. it consists of three subunits of alpha (α), three subunit beta (β), one subunit of gamma (γ), one subunit of sigma (δ) and one subunit epsilon (ε) are tightly bound with F_0 complex.

It has two unit for the synthesis of ATP

- **Rotor (moving unit):** involve- γε and $c_{9\text{-}12}$ ring complex
- **Stator (Stationary unit)** – involve a, b_2 and α_3, β_3, δ complex

Total ATP from oxidation of one molecule of Glucose:

S. NO.	*Metabolic Process*	*Gain of ATP*
1.	Glycolysis = Glucose to pyruvate	7 ATP
2.	Pyruvate oxidation= pyruvate to Acetyl CoA	5 ATP
3.	TCA Cycle = Acetyl Co-A to formation of 6 NADH + 2 $FADH_2$ + 2 ATP	20 ATP
	TOTAL	32 ATP

Inhibitors of ETC (Electron Transport Chain):

S.N.	*Inhibitors*	*Electron Transport Chain*
1.	Rotenone (plant product)	Bind to Complex I and Inhibit to electron transport
2.	Antimycin A (antibiotic)	Bind to complex III and block electron transport
3.	Piericidin A (antibiotic)	Bind to complex I
4.	Amobarbital	Bind to complex I
5.	Cyanide	Bind to complex I
6.	Azide	Bind to complex I
7.	Carbon monoxide	Bind to complex IV and inhibit electron transport between complex IV to O_2

Important Questions:

1. Discuss the TCA cycle and how ATP is generated through ETC (Electron Transport Chain).
2. Discuss the Glycolysis process and how ATP is generated during biotic and Abiotic condition.
3. Write the list of ETC Inhibitors

11 Lipid Metabolism

The production and destruction of lipids in cells is known as lipid metabolism. It involves the formation of phospholipids, glycolipids, sphingolipids, cholesterol, prostaglandins, and other structural and functional lipids that are unique for specific tissues, as well as the breakdown of lipids to meet the body's metabolic needs (such as the production of energy). Both types of fat are used by vertebrates (including humans) to generate energy for the functioning of organs like the heart. Plants also have lipid metabolism; however, the processes are somewhat different from those in animals. The absorption of the fatty acids into the intestinal wall's epithelial cells comes next following the hydrolysis. Fatty acids are packed and transferred to other parts of the body in epithelial cells.

There are some metabolic processes involves Lipid synthesis, Storage, digestion, transport, catabolism, and lipid biosynthesis are all examples of metabolic processes. Beta oxidation, which occurs in the mitochondria and peroxisome cell organelles, is the mechanism responsible for lipid degradation.

Lipid Biosynthesis: Storage Lipids In Adipose Tissues, In Addition To Dietary Fats, Are One Of The Primary Sources Of Energy For Living Things. The Organisms Can Synthesize Cholesterol, Fatty Acid Biosynthesis (Lipid Membranes), And Triacylglycerols Using A Variety Of Mechanisms.

Fatty Acid Biosynthesis

Fatty acid synthesis is a process by which fatty acids are produced from Acetyl-CoA and NADPH by the activity of an enzyme called a fatty acid synthase. The cell's cytoplasm is where this process occurs. The glycolytic process uses carbohydrates as a source of the majority of acetyl-CoA that is transformed into fatty acids. The glycerol needed to create triglycerides (triacylglycerols) with three fatty acids, the end result of the lipogenic process, is also provided by the glycolytic pathway.

The three fatty acids are combined by the ester bond as simple fat. Phospholipid is formed when two fatty acids with glycerol and the third fatty acid with alcohol group through phosphorylated using a group like phosphatidylcholine. Phospholipid is present in the lipid bilayer of plasma membrane that surrounds the organelles with in cell like the Golgi apparatus, mitochondria, endoplasmic reticulum, and cell nucleus etc.

Palmitic acid (saturated fatty acid): Palmitic acid is a type of fatty acid they consist of 16 carbon compounds of saturated fatty acid. Saturated acid is identical in all living systems.

Steps of palmitic acid synthesis: There are seven steps involve given below

Step1: In this step, the carboxylation of acetyl -CoA to form malonyl CoA catalyzed by acetyl- CoA carboxylase

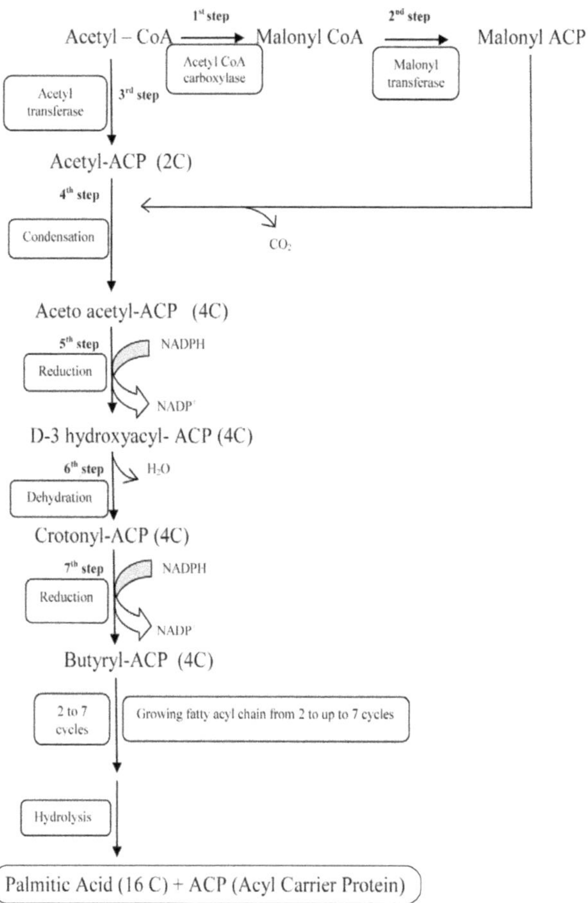

1st step: Acetyl-CoA carboxylase into malonyl CoA catalyzed by acetyl CoA carboxylase.

2nd step: Malonyl CoA transferase into Malonyl ACP. This is catalyzed by malonyl transferase.

3rd step: Acetyl-CoA transferase into Acetyl-ACP. This is catalyzed into Acetyl transferase

Next four step are based on enzymatic reaction. They are including condensation, Reduction, Dehydration and Reduction process. All of four steps are elongation step. Increase chain by addition of two carbon number produce palmitic acid. Entire reaction will be complete up to addition of C15 and C16 chain.

Complete Reaction of Palmitic Acid:

8 Acetyl-CoA + 7 ATP + 14 NADPH + $H_2O \longrightarrow$ Palmitic acid + 7 ADP + 14 NADP+

Beta Oxidation Process

Activation of fatty acid: Each fatty acid is activated by ATP and CoA to generate fatty acyl-CoA. This process is catalyzed by *acyl-CoA synthetase* in cytosol.

Fatty acid transfer from intermembrane space to mitochondria membrane: β (beta) oxidation of fatty acid occurs in mitochondria matrix but synthesis of fatty acid is in cytoplasm. Fatty acids are activated for oxidation start in the cytosol and then transfer into mitochondria for complete oxidation. Fatty acids are long chains that cannot easily transfer into mitochondria Fatty acids are transferred with the help of carnitine (a carrier protein present in the inner mitochondria membrane) if the chain is less than 12 carbons carnitine does not require it.

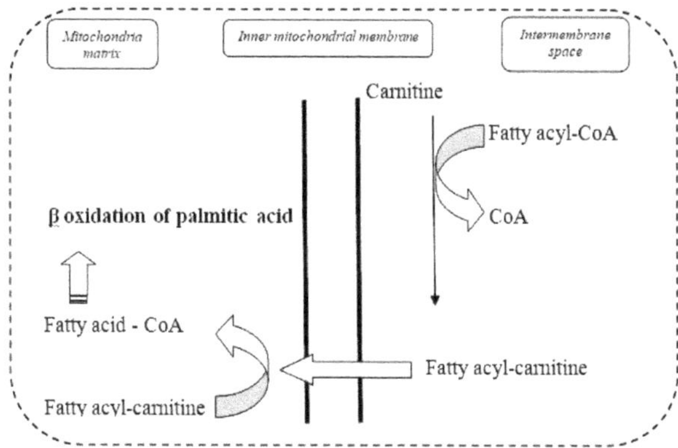

Fig. 4: fatty acid transport from intermembrane to matrix

β oxidation of palmitic acid: Fatty acyl – CoA is long chain of carbon number. fatty acid is oxidized with in within four steps one by one. Two carbon numbers are removed from the chain during this process until the final two carbon fragment is achieved. The reaction of beta oxidation of palmitic acid (a saturated fatty acid) is summarized.

1. Oxidation

$$\text{Acyl-CoA} \xrightarrow[\textit{Acyl-CoA dehydrogenase}]{\text{FAD} \quad \text{FADH}_2} \alpha, \beta\text{- Enoyl-CoA}$$

2. Hydrolysis

α, β- Enoyl-CoA ——— *Enoyl-CoA hydratase* ———→ 3-Hydroxyacyl-CoA

3. Oxidation

3-Hydroxyacyl-CoA ——— NAD^+ → NADH; *Hydroxyacl-CoA dehydrogenase* ———→ β-ketoacyl-CoA

4. Thiolysis

β-ketoacyl-CoA ——— CoA; *Thiolase* ———→ Acyl-CoA + Acetyl-CoA

All four reactions produced a fatty acyl-CoA reduced by two carbons and one molecule of Acetyl-CoA. Acyl-CoA undergoes the beta oxidation cycle once again. When this cycle is repeated using palmitic acid, it produces 8 molecules of acetyl-CoA. The TCA cycle catabolizes Acetyl-CoA, and $FADH_2$ and NADH deliver electrons to the electron transport chain yield ATP.

Complete Oxidation of Palmitic Acid:

Palmitoyl-CoA + 7 CoA + 7 FAD + 7 NAD^+ + 7 H_2O ———→ 8 acetyl-CoA + 7 $FADH_2$ + 7 NADH

Total Yield of ATP:

Reaction	***ATP yield***	***ATP Consumed***
Activation of palmitate to palmitoyl-CoA		-2
Oxidation of acetyl-CoA (8 molecule) by TCA cycle (Total ATP yield in TCA cycle = 10)	8 X 10 = 80	
Oxidation of $FADH_2$ (7 molecule)	7 X 1.5 =10.5	
Oxidation of NADH (7 molecule)	7 X 2.5 = 17.5	
TOTAL ATP =	108-2 = 106	

Note: *Oxidation of 1 molecule of FADH produces 1.5 ATP*

Oxidation of 1 molecule of NADH produces 2.5 ATP

Impotant Questions

1. What is Lipid biosynthesis.
1. Describe four steps of B-oxidation and seven steps of biosynthesis of fatty acid with structure and enzymes.

Multiple Choice Questions

Part -1

1. **Maximum water density occurs**

   (a) 1 °C (b) 2 °C

   (c) 3 °C (d) 4 °C

2. **The term biochemistry was given by**

   (a) Carl Neuberg 1903 (b) Karl Ereky 1919

   (c) Johanson (d) Mendal

3. **How does water liquid at the room temperature**

   (a) Covalent bond

   (b) Non covalent bond

   (c) Hydrogen bond between water molecule

   (d) Van der Waals interaction

4. **Hydrophilic is also known as**

   (a) Polar molecule (b) Non polar molecule

   (c) Amphibolic molecule (d) None of these

5. **Water is a**

   (a) Polar (b) Non polar

   (c) Both (a) and (b) (d) None of these

6. **Buffer maintain to the**

   (a) Acid only

   (b) Base only

   (c) Balance between acid and base

   (d) None of these

7. **pH of buffer solution occurs**

   (a) 7.0 (b) 7.3

   (c) 7.4 (d) 8.0

8. **Carbohydrates are**

   (a) Sugar molecule only (b) Sugar + Protein

   (c) Sugar + Protein + Fat (d) None of these

9. **Monosaccharide sugar molecule are**

   (a) Glucose (b) Fructose

   (c) Galactose (d) All of these

10. **Fructose is a type of**

    (a) Corn sugar (b) Fruit sugar

    (c) Both (a) and (b) (d) None of these

11. **Lactose is a**

    (a) Monosaccharide (b) Disaccharide

    (c) Polysaccharide (d) None of these

12. **Starch is the example of**

    (a) Monosaccharide (b) Disaccharide

    (c) Polysaccharide (d) None of these

13 **Sucrose is a**

   (a) Reducing sugar (b). Non reducing

   (c) Both (a) and (b) (d) None of these

14. **Insulin is a**

    (a) Sugar molecule only

    (b) Hormone only

    (c) both Sugar and Hormone molecule

    (d) None of these

15 **Starch sugar molecule stored in**

   (a) Animal (b) Plant

   (c) Insect (d) Fungus

16. **Sugar transport molecule in plant**

    (a) Starch (b) Sucrose

    (c) Glucose (d) Fructose

17. Lipid in plant as

(a) Energy sugar (b) Protein molecule

(c) Fat molecule (d) All of these

18. Hydrophobic is

(a) Fat soluble (b) Water loving

(c) both Fat and Water (d) None of these

19. Hydrophilic is

(a) Water loving (b) Fat loving

(c) Lipid molecule (d) None of these

20. Palmitic acid is an example of

(a) Saturated fatty acid (b). Unsaturated fatty acid

(c) Both (a) and (b) (d). None of these

21. Soyabean oil , Corn oil and Fish oil are example of

(a) Saturated fatty acid (b) Unsaturated fatty acid

(c) Both (a) and (b) (d) None of these

22. All amino acid are optically active except

(a) Glycine (b) Isoleucine

(c) Proline (d) Alanine

23. Sulphur containing amino acid

(a) Cystine (b) Cysteine

(c) Methionin (d) All of these

24. 21st amino acid is

(a) Selenocysteine (b) Cysteine

(c) Pyrrolysine (d) Glycine

25. Selenocysteine code by

(a) UAA (b) UAG

(c) UGA (d) AUG

26. Stop codon is

(a) UAA, UAG and UGA (b) AUG,GGU and UGG

(c) GUG, AAG and GGA (d) None of these

27. UAG code to

(a) Glycine (b) Alanine

(c) Pyrrolysine (d) Proline

28. Initiation codon is

(a) AUG (b) GUG

(c) UUG (d) All of these

29. Initiation codon in animal mitochondria

(a) AUG (b) AUA

(c) UUG (d) All of these

30. Measurement of 2D structure of protein by

(a) Ramachandran plot (b) Chromatography

(c) SDS PAGE (d) All of these

31. All enzymes are

(a) Protein (b) Amino aci

(c) Lipid (d) None of the

32. Conjugated Enzyme is

(a) Protein part

(b) Non protein par

(c) Both protein + None protein part

(d) All of these

33. Apoenzyme is biological

(a) Inactive protein (b) Active protein

(c) Cofactor (d) None of these

34. Holoenzyme is biological

(a) Inactive protein (b) Active protein

(c) Cofactor (d) None of these

35. lock and key model is related to

(a) Protein (b) Enzyme

(c) Lipid (d) Fatty acyl-CoA

36. Km value is increase in which type of enzyme inhibitor

(a) Competitive inhibition (b) Non competitive inhibition

(c) Uncompetitive inhibition (d) None of these

37. Km value is unaffected in

(a) Competitive inhibition (b) Non competitive inhibition

(c) Uncompetitive inhibition (d) None of these

38. Km value is reduced in

(a) Competitive inhibition (b) Non competitive inhibition

(c) Uncompetitive inhibition (d) None of these

39. Phosphate group is absent in

(a) Nucleoside (b) Nucleotide

(c) Both (d) None of these

40. Thymine nucleotide is absent in

(a) DNA (b) RNA

(C) HNA (d) TNA

41. Uracil nucleotide is absent in

(a) DNA (b) RNA

(c) CeNA (d) TNA

42. Z DNA is a

(a) Right handed helix (b) Left handed helix

(c) both (d) None of these

42. The most common form of DNA

(a) A DNA (b) B DNA

(c) Z DNA (d) HNA

43. Which type of pathway is based on the energy consumption in metabolism

(a) Anabolic process (endergonic)

(b) Catabolic process (exergonic)

(c) Both

(d) None of these

44. Which type of pathway is based on the energy consumption in metabolism

(a) Anabolic process (endergonic)

(b) Catabolic process (exergonic)

(c) Both

(d) None of these

45. Amphibolic pathway considered

(a) Anabolic process (endergonic)

(b) Catabolic process (exergonic)

(c) Both (a) and (b) (d) None of these

46. How much energy is produced in plant glycolysis

(a) 2 ATP (b) 7 ATP

(c) 5 ATP (d) 10 ATP

47. How much ATP produced in pyruvate oxidation

(a) 2 ATP (b) 7 ATP

(c) 5 ATP (d) 10 ATP

48. How much ATP produced in TCA cycle

(a) 2 ATP (b) 7 ATP

(c) 5 ATP (d) 20 ATP

49. How much ATP produced in oxidation of one molecule of glucose

(a) 20 ATP (b) 37 AT

(c) 32 ATP (d) 30 ATP

50. Antimycin A is an inhibitor of

(a) Complex I (b) Complex II

(c) Complex III (d) Complex IV

[Part-1 Answer sheet]

Q.	*Ans.*	*Q.*	*Ans.*	*Q.*	*Ans.*	*Q.*	*Ans.*
1.	(d)	**14.**	(c)	**27.**	(c)	**40.**	(b)
2.	(a)	**15.**	(b)	**28.**	(d)	**41.**	(a)
3.	(c)	**16.**	(b)	**29.**	(b)	**42.**	(b)
4.	(a)	**17.**	(a)	**30.**	(a)	**43.**	(a)
5.	(a)	**18.**	(a)	**31.**	(a)	**44.**	(b)
6.	(c)	**19.**	(a)	**32.**	(c)	**45.**	(c)
7.	(c)	**20.**	(a)	**33.**	(a)	**46.**	(b)
8.	(c)	**21.**	(b)	**34.**	(b)	**47.**	(c)
9.	(d)	**22.**	(a)	**35.**	(b)	**48.**	(d)
10.	(b)	**23.**	(d)	**36.**	(a)	**49.**	(c)
11.	(b)	**24.**	(a)	**37.**	(b)	**50.**	(c)
12.	(c)	**25.**	(c)	**38.**	(c)		
13.	(b)	**26.**	(a)	**39.**	(a)		

Multiple Choice Questions

Part -2

1. **How much ATP produced in oxidation of palmitic acid**
   (a) 108 (b) 106
   (c) 105 (d) 100
2. **How much ATP required in the activation of palmitate**
   (a) One ATP (b) Two ATP
   (c) Three ATP (d) Four ATP
3. **How many steps of beta oxidation of palmitic acid**
   (a) 1 (b) 2
   (c) 3 (d) 4
4. **Lipid are stored in**
   (a) Adipose tissue (b) Liver
   (c) Kidney (d) All of these
5. **Phospholipid is present in the lipid bilayer of**
   (a) Cell wall (b) Plasma membrane
   (c) Golgi body (d) All of these
6. **The fatty acids are combined by which bond**
   (a) Ester bond (b) Hydrogen bond
   (c) Covalent bond (d) All of these
7. **F_1 and F_0 is the subunit of**
   (a) ATP (b) Mitochondria
   (c) Golgi body (d) Cytoplasm

8. Where are ATP formed through electron transport chain

(a) Inner mitochondria membrane

(b) Mitochondria matrix

(c) Golgi body

(d) All of these

9. Electron transport chain are present in

(a) Inner mitochondria membrane space

(b) mitochondria matrix

(c) Golgi body

(d) All of these

10. Glycolysis takes place in

(a) Cytoplasm (b) Mitochondria

(c) Golgi body (d) All of these

11. In the case of eukaryotic, TCA cycle takes place in

(a) Cytosol (b) Mitochondria

(c) Golgi body (d) All of these

12. In the case of prokaryotic, TCA cycle takes place in

(a) Cytosol (b) Mitochondria

(c) Golgi body (d) All of these

13. In the case of eukaryotic, Oxidative phosphorylation takes place in

(a) Plasma Membrane (b) Inner mitochondria membrane

(c) Golgi body (d) All of these

14. In the case of prokaryotic, Oxidative phosphorylation takes place in

(a) Plasma Membrane (b) Inner mitochondria membrane

(c) Golgi body (d) All of these

15. Glycolysis was discovered by

(a) J.C. Bose and Jakub Karol Parnas

(b) Calvin and Gustav Embden,

(c) Krev's and Otto Meyerhof,

(d) Gustav Embden, Otto Meyerhof, and Jakub Karol Parnas.

16. Glycolysis is also known as

(a) Hatch and Slack Pathway

(b) Embden – Meyerhof- Parns (EMP) Pathway

(c) Both (a) and (b)

(d) None of these

17. Which one of the following products is formed during glycolysis

(a) Citric acid (b) Pyruvic acid

(c) Alcohol (d) None of these

18. The fermentation process is

(a) Aerobic process (b) Anaerobic process

(c) Both (a) and (b) (d) None of these

19. The end product of glycolysis

(a) Ethanol (b) Acetyl CoA

(c) Pyruvic acid (d) None of these

20. How many water molecules are released during glycolysis

(a) One (b) Two

(c) Three (d) Four

21. Which of the following is also known as Krebs cycle

(a) TCA cycle (b) EMP pathway

(c) Oxidative pathway (d) All of these

22. Which of the following is a simple lipid

(a) Fatty acid (b) Triacylglycerol

(c) Sterol (d) None of these

23. Which of the following is an essential fatty acid

(a) Oleic acid (b) linolic acid

(c) linolenic acid (d) All of these

24. The general structure of all amino acids are same except

(a) Alanine (b) Proline

(c) Glycine (d) Isoleucine

25. What is the maximum wavelength of tryptophane and tyrosine absorb

(a) 250nm (b) 260nm

(c) 270nm (d) 280nm

26. Which type of bonding is present between secondary protein structure

(a) Van der wall force (b) Hydrophobic Interaction

(c) Covalent bond (d) Hydrogen bonding

27. Which protein structure are stable

(a) Primary structure (b) Secondary structure

(c) Tertiary structure (d) Quaternary structure

28. Which one of the following is a polar amino acid

(a) Cysteine (b) Serine

(c) Glutamine (d) All of these

29. Which one of the following is non polar amino acid

(a) Glycine (b) Alanine

(c) Valine (d) All of these

30. Which of the following vitamins cannot be produced by our body

(a) Vitamin A (b) Vitamin C

(c) Vitamin K (d) All of these

31. Which one of the following fate soluble vitamins is

(a) Vitamin A (b) Vitamin C

(c) Vitamin K (d) Both (a) and (c)

32. Which one of the following water-soluble vitamins is

(a) Vitamin B (b) Vitamin C

(c) Vitamin K (d) Both (a) and (b)

33. Which form of Glucose present in human body

(a) L-form (b) D-form

(c) LD -form (d) None of these

34. Most abundant biomolecule is present on earth surface

(a) Protein (b) lipid

(c) Vitamin (d) Carbohydrates

35. General Formula of Carbohydrates is

(a) $(CH_2O)_n$ (b) $(C_2H_2O)_n$

(c) (CH_2O) (d) None of these

36. The example of an epimer is

(a) Galactose and Mannose (b) Glucose and Galactose

(c) Glucose and Ribose (d) Glucose and Mannose

37. Lactose is a

(a) Monosaccharide (b) Disaccharide

(c) Oligosaccharide (d) polysaccharide

38. Phospholipid and Glycolipid are example of

(a) Membrane lipid (b) Storage lipid

(c) both (d) None of these

39. Triglycerides is the example of

(a) Membrane lipid (b) Storage lipid

(c) both (d) None of these

40. Lipids consists of

(a) Fat, hormone and sterol

(b) fat, phospholipid and sterol

(c) phospholipid, fat and cholesterol

(d) None of these

41. Cholesterol is found in

(a) Plant (b) Animal

(c) Fungai (d) Bacteria

42. Which of the following is the simplest amino acid

(a) Glycine (b) Alanine

(c) Tryptophane (d) Proline

43. Which of the following is the first amino acid in a polypeptide chain

(a) Glycine (b) Glutamate

(c) Selenocysteine (d) Methionine

44. Which of the following is positively charged basic amino acid

(a) Histidine (b) Lysine

(c) Arginine (d) All of these

45. Amino acid has

(a) Amino group (b) Carboxyl group

(c) both (a) and (b) (d) None of these

46. Electron transport chain is present in

(a) Chloroplast (b) Mitochondria matrix

(c) Ribosome (d) Inner mitochondrial membrane space

47. Which ion plays the major role during the final electron transfer to O_2 molecule

(a) Cu (b) Mn

(c) Mg (d) Fe

48. In the TCA cycle, when two molecules of acetyl CoA completely oxidized to CO_2, the total yield of ATP

(a) 10 ATP (b) 15 ATP

(c) 20 ATP (d) 25 ATP

49. Acetyl CoA is a

(a) Two carbon compound (b) Three carbon compounds

(c) Fourth carbon compound (d) None of these

50. Cytochrome c present in

(a) All aerobic organisms (b) Anaerobic organism

(c) Both (a) and (b) (d) None of these

[Part-2 Answer Key]

Q.	*Ans.*	*Q.*	*Ans.*	*Q.*	*Ans.*	*Q.*	*Ans.*
1.	(b)	**14.**	(a)	**27.**	(c)	**40.**	(b)
2.	(b)	**15.**	(d)	**28.**	(d)	**41.**	(b)
3.	(d)	**16.**	(b)	**29.**	(d)	**42.**	(a)
4.	(a)	**17.**	(b)	**30.**	(d)	**43.**	(d)
5.	(b)	**18.**	(b)	**31.**	(d)	**44.**	(d)
6.	(a)	**19.**	(c)	**32.**	(d)	**45.**	(c)
7.	(a)	**20.**	(b)	**33.**	(b)	**46.**	(d)
8.	(b)	**21.**	(a)	**34.**	(d)	**47.**	(a)
9.	(a)	**22.**	(b)	**35.**	(a)	**48.**	(c)
10.	(a)	**23.**	(d)	**36.**	(b)	**49.**	(b)
11.	(b)	**24.**	(b)	**37.**	(a)	**50.**	(a)
12.	(a)	**25.**	(d)	**38.**	(a)		
13.	(b)	**26.**	(d)	**39.**	(b)		

Fill in the Blanks

Part-1

Fill in the Blanks

1. Glycine is the .... amino acid
2. Proline is the .... amino acid
3. ......ATP produced during the TCA cycle
4. Phospholipid is the .... lipid
5. Triglyceride is the .... lipid
6. Final acceptor electron is ....Molecules during the electron transport chain
7. TCA cycle is also known as ....
8. In uncompetitive inhibitors, the km value is ....
9. Cysteine, cystine and methionine is the example of .... amino acid
10. Amino and carboxyl group present in...........
11. .... ATP produced in the oxidation of one glucose molecule
12. NADH carries .... ATP
13. $FADH_2$ carries .... ATP
14. Starch is the example of ....
15. Glucose is an example of ....
16. Lactose is the example of ....
17. Lactose is hydrolyzed into ....and ....
18. Sucrose is hydrolyzed into ....and ....
19. .... ATP are produced during beta-oxidation of palmitic acid
20. .... ATP are produced during aerobic glycolysis

Part-1 Answer Key

1. Simplest
2. Non-optical
3. 20
4. Membrane
5. Storage
6. O_2
7. Krebs cycle
8. Reduced
9. Sulphur containing
10. Protein
11. 32
12. 2.5 ATP
13. 1.5 ATP
14. Polysaccharide
15. Monosaccharide
16. Disaccharide
17. Glucose and galactose
18. Glucose and fructose
19. 106
20. 7

Fill in the Blanks

Part-2

Fill in the Blank

1. AUG code .... amino acid
2. UAA, UAG and UGA are ....
3. UGA code............ amino acid
4. 21th amino acid is ....
5. Sulphur containing amino acid is................
6. Glycolipid is an example of ....
7. Aromatic amino acid is ....
8. .... is storage lipid and.................is membrane lipid.
9. TCA stands for ....
10. ATP stands for ....
11. Oxidation of one molecule of glucose to pyruvate called as..........
12. Glycolysis is also known as....................pathway.
13. End product of glycolysis is ....
14. Pyruvate oxidation takes place..............
15. .... water molecules are used in glycolysis.
16. Antimycin A antibody binds to .... and blocks electron transport.
17. Fermentation takes place ....
18. Convert Acetyl CoA to Succinate during .... Cycle.
19. Chargaff rule is not applicable in ....
20. Diameter of B DNA is..........

Part-2 Answer Key

1. Methionine
2. Stop codon
3. Selenocysteine
4. Selenocysteine
5. Cysteine, cysteine and Methionine.
6. Membrane lipid
7. Phenylalanine, tyrosine and tryptophane
8. Triglyceride, Phospholipid
9. Tricarboxylic acid cycle
10. Adenosine triphosphate
11. Glycolysis
12. Embden-Meyerhof-Parnas (EMP) Pathway
13. Pyruvic acid
14. Mitochondria matrix
15. Two
16. Complex III
17. Cytoplasm
18. Glyoxylate cycle
19. Single-stranded DNA or RNA
20. 20Å

Unit-2

Plant Tissue Culture and Biotechnology

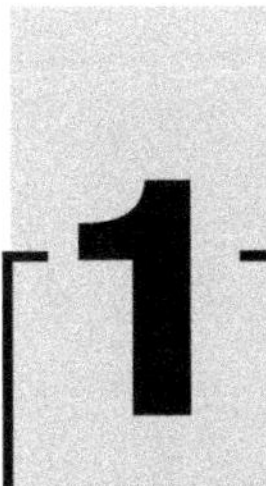

History of Plant Tissue Culture

- **Carrel and Burrows**: Coined Tissue culture term
- **Gottlieb Haberlandt (1898):** Father of plant tissue culture
- **Gottlieb Haberlandt (1902):** Totipotency concept
- **Kotte (1922):** Improve medium
- **Went (1926):** discovered to Auxin
- **Kogl, Haagen-Smit and Erxleben (1933):** IAA (indo-3-acetic acid) isolation
- **Robbins & Maneval (1923):** 20-week maintenance of maize roots using subcultures
- **Gautheret (1932)** : Culture of root tip
- **White (1934**)- Indefinite culture of isolated roots
- **Coplin and Steward (1948):** Discovery of cytokinin
- **Morel and Martin (1952):** Fully developed virus free plant of Dahlia by shoot tip culture and took shoot from virus infected plant
- **Ross Granville Harrison (1970):** Father of animal tissue culture
- **Ross Granville Harrison (1970):** First successful animal tissue culture
- **Guha and Maheshwari (1966):** First haploid plant developed from excised anther of datura innoxia in lab condition
- **Carlson and Smith (1972):** hybrid developed of tobacco from fusion of protoplast
- **Melcher et al (1978):** Pomato (cross b/w tomato and potato)
- **Larkin and Scowcroft (1981):** Proposed somaclonal variation
- **Antony Von Leeuwenhoek (1723):** Living cell
- **Robert Hook (1665):** Cell wall

- **Robert Brown (1851):** Nucleus
- **Fontana (1781):** Nucleolus
- **E. Strasburger (1901):** Plasmodesmata
- **J. E. Purkinje (1840):** Protoplasm
- **Hanstein (1880):** Protoplast
- **J.Q. Plowe (1931):** Cell membrane or plasmalemma
- **Auxin**: Root promoter
- **Cytokinin**: Shoot promoter or cell division
- **Auxin more than Cytokinin in culture medium:** Synthesis of Root
- **Cytokinin more than Auxin in culture medium**: Synthesis of Shoot
- **Auxin and Cytokinin both are equal in medium**: Callus formation
- BAP stand for 6-Benzylaminopurine
- **BAP**: synthesis cytokinin
- Kinetin stand for N6-furfuryladenine
- **Kinetin**: 1st cytokinin synthesis hormone. it was isolated from herring sperm
- **Gibberellins**: Dormancy breaker
- **ABA (Abscisic acid):** Dormancy inducer or Anti gibberellins
- **Ethylene**: Fruit ripening
- **CCC(Cycocel):** Growth retardant or Prevents
- **Mallic Hydrazide**: Also called as sprout stop in onion
- **IAA:** Natural plant hormone
- **24D:** Synthetic plant hormone (Herbicide hormone <20 ppm)
- **Dicamba**: Active herbicide plant hormone
- **Murashig and Skoog (1962):** Discovered M.S. Medium (mostely used for organogenesis and micropropagation, callus culture)
- **Linsmaier and skoog (1965):** LS Medium (micropropagation and cell suspension culture)
- **O.L. Gamborg (1968):** Gamborg (B5) Medium (used for protoplast culture)
- **J. P. Nitsch (1969):** Nitsch and Nitsch (NN) Medium (used for anther culture and pollen culture)
- **P.R. White (1963):** White's Medium (used for root culture, shoot culture, callus culture)

- **Hanning (1904):** First embryo culture
- **Steward et al., (1958):** Discovery of somatic embryogenesis
- **Cocking (1960):** Protoplast isolation
- **San Noem (1976):** Ovary culture
- **Morel and Martin (1960):** Clonal propagation
- Callus induction from explant
- Explant source- seed, leaves, root, apical meristem etc.
- **Explant size**: 1.0 to 1.5 cm

2 Tissue Culture Medium

Tissue culture media is a type of source in which plants receive nutrients for healthy growth. Culture media is a combination of different components like macronutrients, micronutrients, vitamins, amino acids, gelling agent etc.

There are five types of nutrients media with year and specificity:

S.No.	*Name of media*	*Year*	*Specificity*
1.	Murashige and Skoog (MS) Media	1962	Organogenesis (callus culture), micropropagation, single cell culture
2.	White Media	1963	Root culture
3.	Linsmaier and Skoog (LS) Media	1965	Micropropagation, organogenesis, callus culture, cell suspension culture
4.	Gamborg (B5) Media	1968	Protoplast culture
5.	Nitsch and Nitsch (NN) Media	1969	Haploid culture (anther/pollen culture)

Component of Nutrient in media: There are several components that are given below

- Macronutrient:
  - Basic nutrient (C, H, O), Primary nutrient (N, P, K), Secondary nutrient (Ca, Mg, S)
  - Requirements: - > 0.5
- Micronutrients:
  - Mn, Fe, Zn, Cu, B, Ni, Cl and Mo
  - Requirements: - < 0.5
- Plant Growth Hormone:
  - If Auxin/cytokine high: Root Formation
  - If Cytokinin/Auxin: Shoot
  - If both are equal (auxin and cytokinin): formation of callus

- **Note:** *callus- undifferentiated mass of cell*
- GA_3:
  - Inhibit adventitious root and shoot
  - Breakdown seed dormancy
- **ABA**:
  - Anti gibberellin
  - Promote somatic embryogenesis and organogenesis

- Carbone and nitrogen source: sucrose mostly and Nitrate mostly
- Gelling agent:
  - Agar:
    - Requirements: - (0.6 to 0.8%)
    - Agar obtained from red algae *(Gelidium amansii)*
  - Agarose:
    - Obtained from purifying agar
    - Required for: cell suspension culture, protoplast culture
  - Alginate:
    - Protoplast culture (Mostly)
- Amino acid:
  - Glycine (Neutral amino acid) mostly used
- Vitamins: B (B_1, B_2, B_5, B_6) C, D, and E mostly used

Important Questions:

1. What is Plant Tissue culture Media and its composition.
2. Write the list of specific media using in Plant Tissue culture.
3. What are gelling agents.

3 Plant Tissue Culture

Plant tissue culture is the process of growing plant seeds, organs, explants, tissues, cells, or protoplasts on a synthetic nutritive medium in sterile, temperature- and humidity-controlled conditions.

There are Some Terms which are Related to the Plant Tissue Culture:

- **Totipotency**: Ability of single plant produced to the whole plant called totipotency.
- **Explant**: Any part of plants which is used for the culture.
- **Callus**: undifferentiated mass of cell.
- **Dedifferentiation**: Meristem cells produced to the callus.
- **Redifferentiation**: Callus produced to whole plants.
- **Regeneration**: formation of new organ from culture of explant.
- **Adventitious**: shoot or root develop from unusual points.
- **Embryo rescue**: isolate embryo during stress condition of plants for the saving purpose
- **Apical meristem**: top portion of plants.
- **Xylem**: it is non-living part except parenchyma and responsible for water and nutrition transport.
- **Phloem**: phloem is living part except fiber and responsible for food and other organic material.

Important Questions:

1. What is Totipotency?
2. What is Callus?
3. What is Embryo Rescue?
4. What is Explant and its size in Micro-propagation?

4 Sterilization and Explant Culture

Sterilization: A process that uses chemical or physical methods to remove all types of microbial life.

S. No.	*Type of Sterilization*	*Uses*
1.	**Physical sterilization**	
a.	Autoclave	121°C at 15 psi (Pound per square inch) for 20 minutes
b.	Hot air oven	It is used for apparatus sterilization and drying
2.	**Chemical sterilization**	
a.	Ethanol	70% used for surface sterilization and 95% flame sterilization
b.	Sodium Hypochlorite NaClO	1% used for surface sterilization of explant
c.	Calcium Hypochlorite $Ca(ClO)_2$	2.5% used for 5 min. in surface sterilization of explant
d.	Mercuric chlorite ($HgCl_2$)	0.1% used for 8 min. in surface sterilization of explant
e.	Silver Nitrate ($AgNO_3$)	0.1%-1.0% for 3-5 min. in surface sterilization of explant
f.	Mercuric chlorite	0.01% to 1% used for surface sterilization of explant
g.	Washing sterilization	Apparatus washed by Teepol (nontoxic detergent)

Important Questions

1. What is Autoclave?
2. What is Sodium hypochlorite?
3. What is Sterilization and its classification?

5 Triploid Production

Diploid Species

- One sperm fertilizes to the egg cell generating the diploid zygote (2n)
- Normal condition
- single fertilization
- Seed formation

Triploid Species

- Also called endosperm (3n)
- One sperm fertilizes to the egg cell and other sperm fertilizes the central cell giving rise to endosperm (3n)
- Triploid is a double fertilization process
- Seedless (No seed formation)
- **Example**: Banana, watermelon (Pusa Bedana), Mulberry etc.

Endosperm culture: Endosperm culture is a technique used in plant tissue culture to regenerate plants from the endosperm tissue of seeds. The endosperm is the nutrient-rich tissue surrounding the embryo in a seed.

History

- Lampe and Mills (1933) were the first to report the proliferation of immature endosperm tissue in maize cultivated on medium containing potato extract.
- **Nakano *et al.* (1975):** Cultured immature rice endosperm and achieved organogenesis.
- Rangaswami and Rao (1963) developed a mature endosperm of Santalum album (Chandan) up to callus successfully.

Process in *In-vitro*

The process begins by collecting mature seeds and sterilizing them to remove any surface contaminants. The endosperm is carefully extracted from the seed, avoiding damage to the embryo. The isolated endosperm is then placed onto a culture medium containing a combination of nutrients, growth regulators, and other necessary components.

Under controlled laboratory conditions, the endosperm tissue starts to divide and form a callus, a mass of undifferentiated cells. This callus can be further stimulated to differentiate and develop into shoots, roots, and ultimately, complete plantlets. The culture conditions, including the composition of the medium and the growth regulators used, are crucial in determining the success of endosperm culture.

Applications

Endosperm culture has various applications in plant research and breeding. It allows for the propagation of valuable plant species that are difficult to regenerate through conventional means, such as orchids or certain fruit trees. It also provides a means to study the development and physiology of the endosperm tissue itself.

Important Questions

1. What is difference between Diploid and Triploid species?
2. What is Endosperm culture ?
3. What is the example of Triploid species ?

6 Haploid Culture

Haploid culture is one of the *in-vitro* culture in which we can used haploid cells (anther, pollen, ovule, ovary and floral bud) for culture to develop whole plant (n) under aseptic condition.

Approaches: There are two approaches for haploid culture

- Androgenesis
- Gynogenesis

Explant for the Haploid Culture

- **In the case of androgenesis**: Explants are in the form ovary, unpollinated ovule and floral buds.
- **In the case of gynogenesis**: Explant are in the form anther and pollen

Culture Media for the Haploid Culture

Nitsch and Nitsch (NN) Media: J.P. Nitsch created the media in 1969 to grow an in-vitro anther culture of Nicotiana from the Solanaceae family. This media, which is primarily utilised in the context of anther culture, refers to haploid culture.

Type/Source of Haploid

(a) Natural: There are natural phenomenon

- Parthenogenesis, apomixis, apogamy and apospory

(b) Artificial (In-vitro technique of plant tissue culture)

- Gametic embryogenesis culture: androgenesis and gynogenesis

1. Androgenesis (Anther/pollen)

Androgenesis is the in-vitro creation of a haploid plant from an anther and a totipotent pollen grain by cell division and differentiation.

There are two kinds of it.

Direct androgenesis and Indirect androgenesis

- **Direct androgenesis:** The microspores behave like a zygote and undergoes to form embryoid which ultimately give rich to plantlets.
- **Indirect androgenesis** occurs when microspores repeatedly divide to generate a callus tissue that develops into haploid plantlets.

1.1 Mechanism of Anther Culture: There are several steps of anther culture mechanism in which steps are include from explant isolation to up to developed whole plants (haploid plants).

Step1: Selection, isolation and collection of unopened flower bud (Explant)

⇓

Step 2: Surface sterilization with alcohol (70% ethanol)

⇓

Step 3: Isolate anther from flower bud and kept separately

⇓

Step 3: Acetocarmine test in which anther selected in the form first meiotic division

⇓

Step4: Inoculated of anther in medium in which containing glutamine, L-serine and inositol

⇓

Step 5: Incubate the culture at 250 C for 15-day, anther is grown in the form of embryoid

⇓

Step 6: Embryoid transfer on rooting medium became plantlets after 4 - 5 weeks after then plant transfer to greenhouse

1.2 Mechanism of Pollen Culture: There are several steps of anther culture mechanism in which steps are included from explant isolation to developed whole plants (haploid plants).

Step1: Collect anther from flower bud and pollen grain are isolate

⇓

Step 2: Pollen Placed in beaker who are sterile and in which contain 20ML growth medium (M S Medium)

⇓

Step 3: pollen filtered through nylon sieve

⇓

Step 4: The filtered or pollen suspension is then centrifuged for 5 minutes at low speed (500-800 rpm). The supernatant is discarded, and the pollen pellet is suspended in fresh liquid media before being cleaned and resuspended in fresh liquid medium.

⇓

Step 5: Take 2.5 ml pollen suspension in pipette and spread in Petri dish. Pollen is best grown in liquid medium

⇓

Step 6: Petri dish are incubated at 30oC under low intensity of white cool light

⇓

Step 7: observe young embryoid after 28 to 30 days

⇓

Step 8: Develop haploid plantlets and then transfer to greenhouse

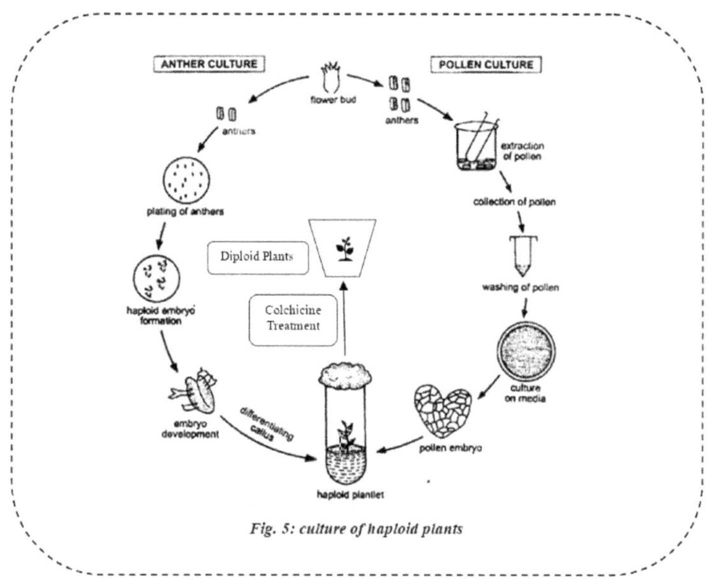

Fig. 5: culture of haploid plants

Fig. 5: culture of haploid plants

2. Gynogenesis (Ovary, Ovule or Flower Bud)

Ovary or ovule culture that results in the creation of gynogenic haploids. Ovarian or ovule culture can be used to create haploid plants. Female angiosperm gametophytes can be induced to develop into sporophytes. The plants that result are known as gynogenic haploids.

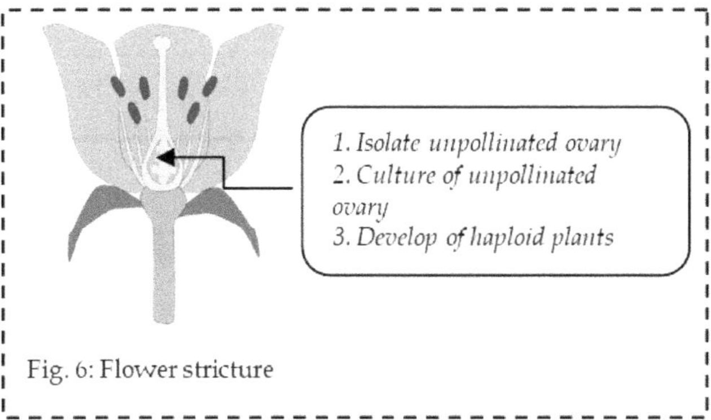

Fig. 6: Flower structure

San Noem 1976 created the first gynogenic haploid from hordeum vulgare ovary culture.

Process of Culture:

Step 1: Isolation of unpollinated ovaries from the flower

⇓

Step 2: Removal of unnecessary parts from the flower like calyx, corolla and stamens

⇓

Step 3: Surface sterilization of ovary from alcohol

⇓

Step 4: Inoculation of ovary in a suitable medium

⇓

Step 4: Incubation of culture

⇓

Step 5: Formation of callus /embryo

⇓

Step 6: Development of Haploid Plan

Application

- For mutation study
- For plant breeding purpose
- For secondary metabolites production
- Haploid plant are used in molecular biology and genetic engineering

Important Questions

1. What is Haploid culture?
2. What is Androgenesis process and describe the anther and pollon culture steps also?

7

Embryo Culture

Embryo Culture

- Embryo culture is the technique of cultivating embryos from seeds and ovules on nutritive media.
- Somatic embryogenesis and regeneration process carried to embryo culture

Embryo Culture are Two Types: immature and Mature Embryo Culture

Immature culture: The globular and heart-shaped developmental stages are ideal for the culture. Such cultures are managed to meet the nutritional needs of growing embryos. This type of culture is required for the rescue of the embryos.

Mature embryo: in this culture, the mature seed are required and break dormancy of seed during culture. Several portions of the developed seed embryo can be removed surgically. These embryo segments are cultured to examine how various embryonic components relate to the embryo's final form in culture.

Some growth regulators are required for the embryo culture:

- ABA does require for the maintain osmatic balance
- Auxin does not require in embryo culture
- GA_3 does required for the breaking seed dormancy

Embryo Rescue: It was one of the first successful in vitro/tissue cultivation methods. Embryo rescue is a procedure used to protect and encourage the development of weak, immature, or hybrid embryo into a whole plantOne of the most important strategies in plant breeding is embryo rescue. It enables the development of weak embryos that have been carefully prepared through interspecific or intergeneric crossings into whole plants.

Why does Embryo Rescue?

Embryos that arise from interspecific or intergeneric crosses are frequently undeveloped and unable to develop into plants in nature, and in many cases where crosses are conducted between incompatible plants, the embryos are aborted. Tissue culture procedures, such as embryo culture, save these plants and help to generate hybrid plants.

Advantages

- Most of suitable stages for embryo isolation for culture: Heart shaped stage
- Ovule and ovary also be cultured
- Recovery of haploid plants
- Recovery of distant hybrids
- Overcoming dormancy and short breeding cycle

Important Questions:

1. What is Embyo culture and describe the Mature and Immature Embryo culture also?
2. Why does Embryo rescue and its advantages?

8 Micropropagation

- Micropropagation is a type of plant tissue culture in which we use modern technology to rapidly multiply stock plant material to produce a huge number of young plants.
- Also called clonal propagation
- **Clone**: clone is a type of asexual production
- Ist clonal propagation was orchid propagation developed by **Morel (1960)**
- Apical meristem is required as an explant during micropropagation
- Virus free plants are developed through the culture of apical meristem as an explant
- Dehlia was first virus free plant developed during apical meristem culture by Morel and Martin in 1952

Steps of Micropropagation

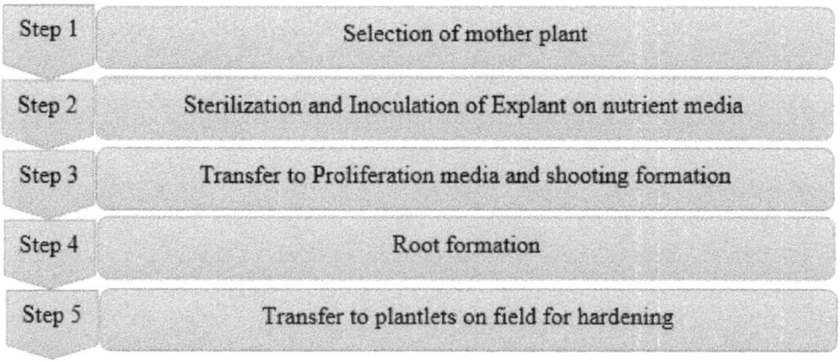

Step	
Step 1	Selection of mother plant
Step 2	Sterilization and Inoculation of Explant on nutrient media
Step 3	Transfer to Proliferation media and shooting formation
Step 4	Root formation
Step 5	Transfer to plantlets on field for hardening

Step 1: Selection of Mother Plant

- Fresh plants should be selected for the isolation of desired explant for the culture
- An axillary/ Apical meristem as explant selected for the culture

Step 2: Sterilization and Inoculation of Explant on Nutrient Media

- The explant should be sterilized from alcohol (70% ethanol)
- After the sterilization process, you can put on the culture medium called as explant inoculation

Step 3: Transfer to Proliferation Media and Shooting Formation

- After inoculation, explant is proliferated and formation of shoot
- Cytokinin (kinetin or BPA) are play a major role in the formation of shoot

Step 4: Root Formation

- After shoot formation, fresh medium is required for the formation of root
- Subculturing for root formation
- Auxin plays a major role in root formation develop plantlets

Step 5: Acclimatization (Hardening)

- Plantlets transfer to the greenhouse or field
- Developed virus free plants through the tissue culture process

b. Organogenesis

Callus differentiate into two major parts root and shoot called organogenesis. Skoog and Miller 1957 reported regulation of organogenesis due to hormonal balance (auxin and cytokinin). If auxin will be high in medium than cytokinin formation of root whereas cytokinin will be in medium than auxin formation of shoot.

Caulogenesis: Regeneration of shoot from callus

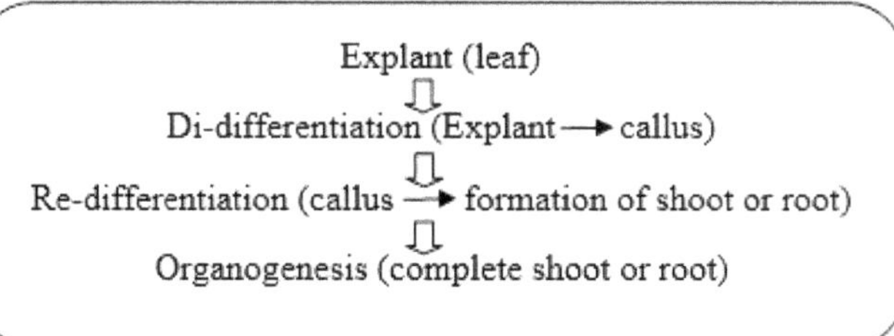

Rhizogenesis: Formation of root from callus

Explant → Callus → Regeneration of shoot

Direct organogenesis: Formation of shoot or root without callus

Explant → Callus → Formation of root

Indirect organogenesis: Formation of shoot or root through callus

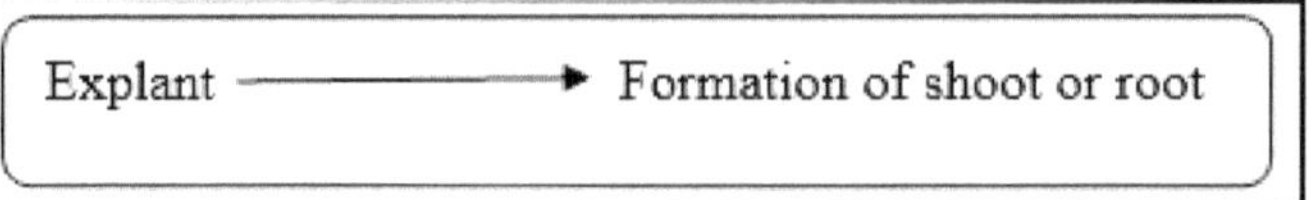

c. Somatic Embryogenesis

- Somatic embryogenesis is an artificial process that results in the formation of a plant or embryo from a single somatic cell. Somatic embryos are created from plant cells that do not normally participate in embryo development, i.e. normal plant tissue. A somatic embryo is not surrounded by endosperm or seed coat.
- The first somatic embryogenesis plant was carrot developed by Renert and Steward *et al.,* 1958
- Developed bipolar structure by non-zygotic somatic cell
- Somatic embryogenesis is an asexual process through plant tissue culture under *in-vitro* condition.
- Auxin (2, 4D) plays a major role in the formation of somatic embryogenesis
- ABA plays a major role during maturation and development of somatic embryo

Indirect somatic embryogenesis: when explant produced callus then callus into mature embryo

Direct somatic embryogenesis: Embryo produced from explant without callus formation

Explant → Callus → Embryonic clump → Mature somatic embryo

Artificial seed: An artificial seed (also known as a synthetic seed) is a package containing plant structures such as somatic embryos, buds, shoots, or other meristematic tissues that can be sown in the same way as a traditional seed to grow a new plant.

General Procedure for the Synthesis of Artificial Seed:

- Culture of somatic embryo
- Shoot bud formation and embryo maturation
- Encapsulating matrix includes calcium alginate, agar and carrageenin's
- Green house planting and mass production

Methods:

1. Hydrated Method:

- In this method, several water-soluble agents such as calcium alginate, gel rite, locust bean gum, and sodium alginate with gelatin are employed in this approach. However, alginate is the best agent to use.
- **Radenbergh *et al.*,** (1986) developed hydrated method mixing somatic embryos with sodium-alginate and then putting them into a $CaCl_2$ solution to form calcium alginate beads.
- Around 30-55% of the embryos were capable of germinating and producing seedlings.

2. Desiccated Method:

- Encapsulation of somatic embryos is followed by desiccation in a laminar hood.
- It was developed by **Kin and Janick (1989)** used synthetic seed coverings to create dehydrated artificial seeds.
- A similar volume of embryo suspension was mixed with a 5% solution of polyethylene oxide.

Important Questions:

1. What are Micropropagation and define the different steps also?
2. Describe the Organogenesis process in detail?
3. What are Somatic Embyrogenesis?
4. Describe the Artificial seed and its methods?

9 Somatic Hybridization

- Somatic hybridization is the process of fusing protoplasts to form a hybrid cell and then cultivating the hybrid cell to create a hybrid plant.
- Somatic hybridization was first introduced by **Carlson** *et al.* in Nicotiana glauca.
- First man made hybrid – **Thomas Fairchild 1717**

Protoplast Isolation

- Plant cells are without cell wall called protoplast
- Protoplast term given by **Hanstein 1880**
- First Mechanical Protoplast isolation in cucumber and beets was given by **Klerker *et al.* 1892**
- first enzymatic protoplast isolation in tomato was given by **Cocking *et al.* 1960**
- First enzyme was used in first protoplast isolation: cellulose (hydrolytic enzyme) from tomato root tips in 1960.
- First hybrid plant developed by protoplast fusion: **Tobacco**
- Optimum pH of protoplast isolation: 4.5 to 6.5
- Optimum pH of protoplast culture: 5.5 to 6.0
- Explant for protoplast culture: leaf or Embryonic cell
- Difference between protoplast and protoplast:
  - Protoplast – cell without cell wall
  - Protoplasm- cytoplasm (cell organelles) + nucleus

Sevel Enzymes used for the protoplast isolation

S. No.	Enzymes	Sources
1.	Cellulose	Trichoderma viridae
2.	Hemicellulose	Aspergillus sp.
3.	Meicelase	Trichoderma viridae
4.	Pectinase	Aspergillus niger
5.	Pectolyase	Aspergillus japonicum
6.	Zymolyase	Arthobacter

2. Protoplast Fusion and Cybrid

- It is a man-made process
- An approach for the production of hybrid or cybrid and overcoming species barrier
- Fusion of two cytoplasm together with the help of enzymatic reaction
- PEG (poly ethylene glycol) is used for fusion of protoplast
- Optimum pH for fusion of protoplast: 10.5
- Ca^{2+} are involve in protoplast fusion

Method of Protoplast fusion

PEG Mediated: High pH (10.5) + Ca^{2+} + PEG

Somatic Hybridization (Hybrid) and Cybridization (Cybrid)

Somatic Hybridization

- Two protoplasts fused together through PEG (polyethylene glycol)
- It developed by nuclear fusion

Cybridization (Cybrid)

- Fusion of one protoplast with enucleated protoplast (without nucleus) together through PEG
- Cybrid developed from fusion of cytoplasm without nuclear fusion.
- Nucleus from one parent and cytoplasm from both parent
- Mitochondria and Chloroplast genome from both parent
- Cybrid also called cytoplasmic hybrid or cybridization

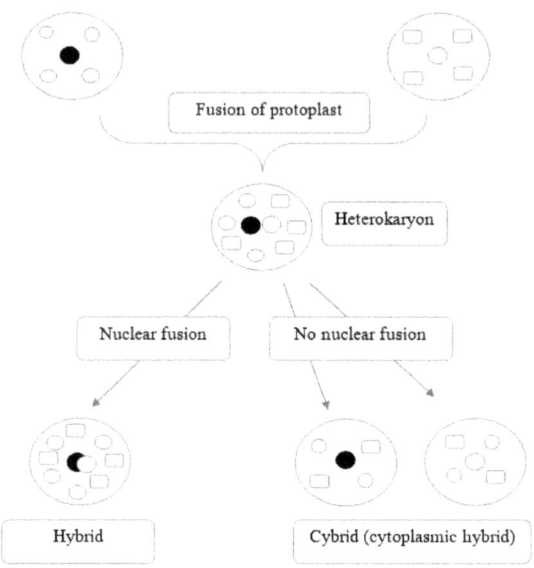

Fig. 7: Development of cybrid and hybrid through the protoplast fusion

Important Questions:

1. Describe the somatic hybridization .
2. Define the protoplast isolation process and some specific enzymes using protoplast isolation also.
3. Describe the protoplast fusion and cybrid in details.

10 Secondary Metabolite Production

- Secondary metabolite is a natural plant product; however, it is not directly involved in plant growth and development.
- It produced from outcome of primary metabolism
- It depends on secondary biosynthesis pathway
- Secondary metabolite term coined by Albrecht kossel
- Plant metabolism discovered by: F.W. Serturner 1880

Group of Secondary Metabolites:

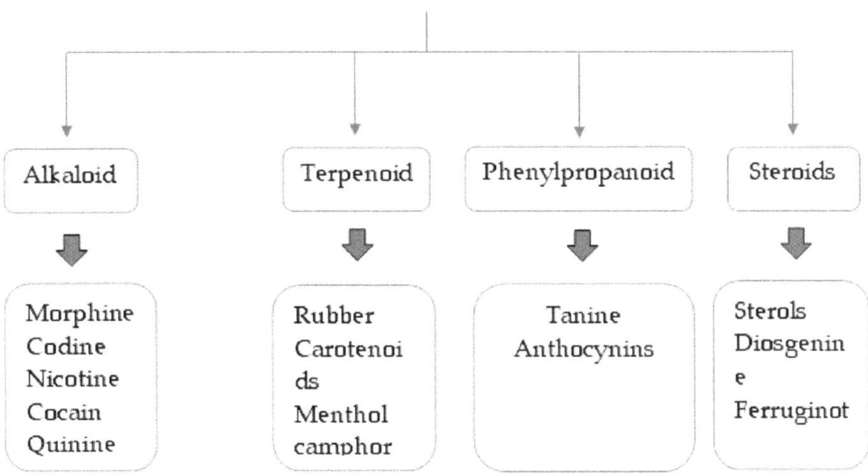

- DMSO (Permeabilizing agent) promote to secondary metabolite.
- Cell suspension culture is used for the Secondary metabolite production.
- Hight level production: late exponential phase.

- Low level production: lag phase (beginning phase).
- Organic compounds defined as secondary metabolites are the chemical structures responsible for biological activity in medicinal plants.

Important Questions

1. What is Secondary Metabolite?
2. Describe the classification of secondary Metabolite production.

11 Cell Suspension Culture

- Also called single cell culture.
- Totipotency: The ability of a single cell to developed whole plants called totipotency.
- Tissue and cells are grown in liquid medium, resulting in a suspension of single cells and clumps of cells.
- The principal method for researching plant cell secondary metabolism is cell suspension culture.
- Cell suspension culture required the optimization of the cell line, cultured medium, and bioreactor system.

History

- **W. H. Muir (1953):** first reported callus fragments could be culture in the form of callus.
- **Nickle (1956):** first reported *phaseolus vulgaris* of a continuously maintained cell suspension culture.

Steps of Cell Suspension Culture

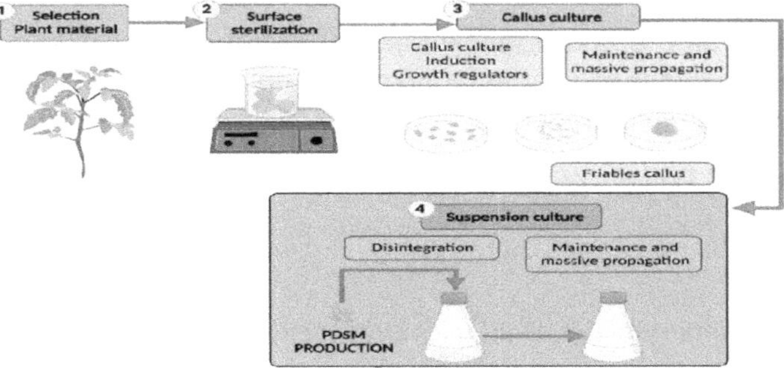

Fig.-8. *In-vitro* cell suspension culture, PDSM: – Plant Derived Secondary Metabolites. Link: https://doi.org/10.3390/plants10122762

Type of Cell Suspension Culture: There two type of suspension culture

1. Batch Culture

- Batch culture is a type of cell suspension culture in which cells grow in a limited volume of agitated liquid media.
- These cultures are kept maintained continuously by subculturing.
- Batch cultures are most often maintained in conical flasks shaken at 80 to 120 rpm on orbital platforms.
- It is a close type of culture. During the incubation period, no nutrients or waste materials are added or removed.
- Culture in a fixed volume of liquid medium
- **Sigmoid (S-shaped)** growth pattern

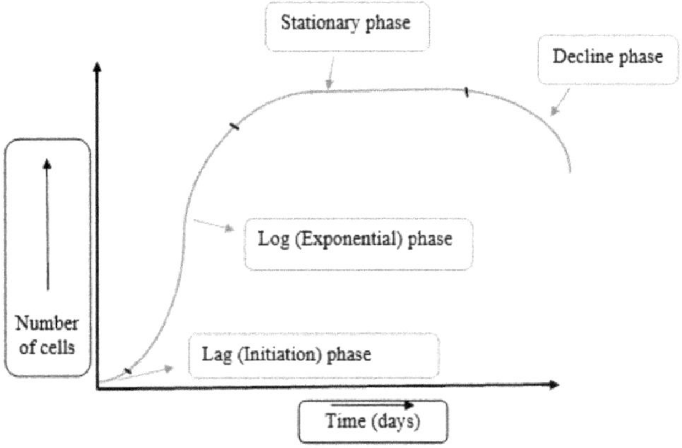

Fig.9: Growth curve of Batch Culture

- **Lag phase**: Initiation stage of batch culture. its cells adjust to the environment and prepare for division.
- **Log phase:** Increase the number of cells by using the growth medium.
- **Stationary phase**: constant (stop) cell growth due to exhaustion of growth factor or accumulation of certain toxic metabolite in medium after 3 to 4 generation called stationary phase.
- **Decline phase**: some cell that existed in stationary phase lead to death phase.

Note:

- *At stationary phase, need to change medium.*
- *Overall batch culture is harmful process for the study of growth and metabolism of cell.*

Other Batch Cultures Include

- Slow rotatory culture
- Shaker culture
- Spinning culture
- Stirred culture

2. Continuous Culture

- The old liquid medium is continuously replaced by fresh liquid medium in a continuous culture system to stabilize the physiological states of the growing cell. Nutrient depletion does not occur in this system due to the continuous input of nutrients, and the cell is always in the steady growth phase.
- Growth pattern occurs **Parabolic/hyperbolic**

Type of continuous culture: Continuous cultivation is classified into two categories.

- In closed type
- In open type

A. In Close Type

In a closed continuous culture system, both the supply and removal of culture fluid are carefully controlled. Fresh nutrient medium is continuously added at a controlled rate, and an equal volume of spent medium, containing waste products and cells, is simultaneously removed from the bioreactor.

B. In Open Types

In an open continuous culture system, culture fluid (usually cell suspension medium) is continuously supplied to the bioreactor, but there is no active control or removal of waste products.

It is two types:

- **Chemostate**: Add the fresh medium for maintains steady cell growth.
- **Turbidostate**: The biomass of the cells is always kept at optimal yield. In contrast to chemostat culture, you can feed all nutrients in excess. No new medium should be added.

Important Questions

1. Describe the cell susponsion culture and its classification with figures.
2. Close continuous and open continous culture.
3. Describe the Batch culture and its different Phages

Somaclonal Variation and Genetic Fidelity

- Genetic variation developed in plants through the plant tissue culture under in vitro condition. It can be detected as genetic and phenotypic traits.
- Somaclonal variation term given by **Larkin and Scowkraft (1981)**
- Sudden change the genetic material due to In-vitro culture of somatic cells produced desirable changes in yield increase, disease resistance and quality improvements
- Create the variation in production of clone by culture of somatic tissues.
- Adventitious proliferation cause somaclonal variation
- Somaclonal variation occurs during micropropagation
- Desirable feature obtained by somaclonal variation like crop yield improvements, Disease resistance etc.

Classification of Somaclonal Variation: Somaclonal variation derived from somatic cell through plant tissue culture.

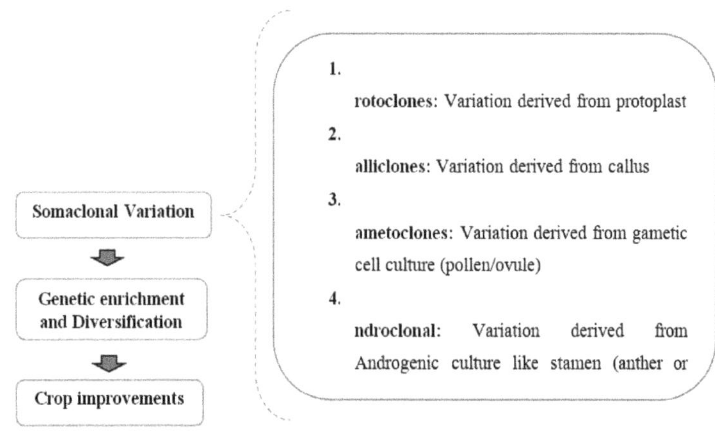

Steps Involve in Somaclonal Variation:

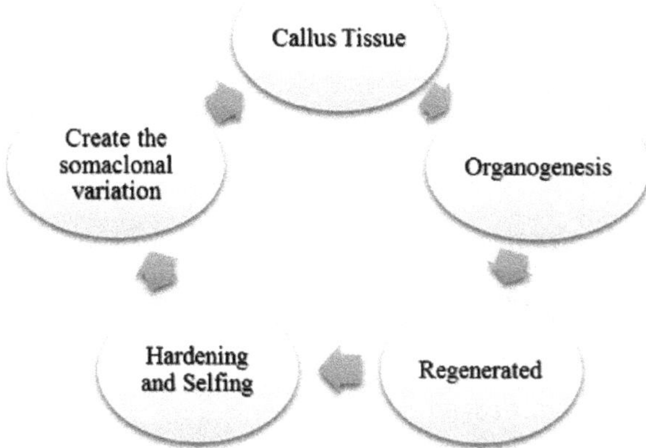

Source of Variation:

- **Physical sources**: Temperature
- **Chemical sources**: pH, plant growth hormone, Culture media and media composition

Variation Shows at:

- Growth pattern
- Regeneration pattern
- Phenotype
- Genotype (ploidy level)

Determination Technique of Somaclonal Variation:

Genetic fidelity: As somaclonal variants continue to be a key barrier in plant multiplication by tissue culture, genetic fidelity testing of tissue culture grown plantlets has gained importance. **(Bairu *et al.,* 2011; Rizvi *et al.,* 2012).** it determines to the Somacolnal variation with the help of the molecular marker like RAPD and SSR.

Molecular Variation Level:

At genetic level:

- Mutation in chromosome abbreviation (addition, deletion, inversion, translocation)
- Point mutation
- Frameshift mutation
- Transposons

At Epigenetic Level

- Mutation in Histone protein
- Methylation
- Acetylation

At Biochemical Level

- Carotenoid pathway
- Nitrogen metabolism
- Antibiotic resistance

Advantages of Somaclonal Variation

- Increased and improved production
- Improved secondary metabolites
- Developments of resistant plants
- Create the additional genetic variation
- Help in crop improvements

Disadvantages

- Leads to undesirable results
- Genetically unstable product
- Develop pleiotropic effect
- Extended field trails

List of Somaclonal variety: There are some somaclonal variety which developed from plant tissue culture.

S. No.	*Somaclonal varieties*	*Sources*	*Improved Traits*
1.	Ono	Sugarcane	Fiji disease resistant
2.	Pusa jai Kisan	Mustered	Increase Yield and oil content
3.	Scarlet	Seet potato	Skin color

Important Questions:

1. Describe the somaclonal variation and its different steps also.
2. Describe the genetic fidelity in details.
3. Write the list of somaclonal varieties.

13 Germplasm Conservation

The major difference between conservation and preservation

CONSERVATION	PRESERVATION
Preservation of the natural resource in the area. like land, forests etc.	Keeping something from harm / damage. Like foods, pickles etc.

Conservation

- One of the key areas that knowledge needs to continually improve and revolutionize is plant genetic conservation.
- The phenomena of conservation aids in preserving the genetic foundation required for breeding.
- This makes it possible to grow and choose crop types with favorable traits that can later be applied to the food, fuel, and health sectors. The conservation is involved in some points:
  - Preserve to Breeding line.
  - Conserve to important species for commercial purposes.
  - Conservation stock for genetic.
  - Use of wild species.

Germplasm Conservation

- The term "germplasm" refers to the plant's genetic materials, such as tissues or cells, which are being conserved in order to develop desirable breeding traits. Gene banks, plants grown in nurseries, and laboratory culture are the sources of these resources. Typically, wild species and genes that are intended to capture characteristics of plants as a result of natural selection are included in collections of germplasm.

- Breeders use plant germplasm as a general resource to create new cultivars. This involves pollen, cultured cells, seeds, leaves, and stems. As a result, it supplies the essential raw material for creating plant varieties with high commercial value.

Requirements for germplasm Conservation: There are several reasons why breeders use this technique which is as follows:

- Loss of genetic diversity among plant species
- Micro propagated
- For seedless plant
- Genetic stock

Approaches for Germplasm Conservation

***In-situ* conservation :** In-situ conservation is the preservation of germplasm in its native habitat through the creation of biosphere reserves (national parks/gene sanctuaries).

Ex-situ Conservation

- The main technique for preserving germplasm derived from cultivated and wild plant materials is ex-situ conservation.
- Under the right circumstances, the genetic material from plant cells, tissues, or organs grown in a lab can be stored as gene banks for long-term storage.
- For the construction of gene banks to be successful, it is crucial to have a thorough understanding of the genetic makeup of plant populations as well as the procedures used for sampling, regeneration, maintaining gene pools, etc.

Germplasm Conservation in the Form of Seeds

- Seeds are the most common and practical material for preserving plant germplasm.
- Many plants are propagated through seeds, and seeds take up very little space. It is conveniently transportable to many locations.

Limitations

- Seed viability gets reduced or lost.
- Seeds are susceptible to insect or pathogen attack, which frequently results in their demise.
- This method is only applicable to seed propagating plants and has no application to vegetatively propagated plants such as potato, Ipomoea, and so on.
- Clone maintenance through seed preservation is challenging.

In vitro Methods for Germplasm Conservation

In vitro methods employing shoots, Meristems and Embryos are ideally able for the conservation of germplasm of vegetatively propagated plants. The plants, seeds and genetically engineered materials can also be preserved by this in vitro approach.

Cryopreservation

The term "cryopreservation" refers to preservation in a frozen state. Storage of germplasm at extremely low temperatures is known as cryopreservation. Cryopreservation can be done in various ways

- Over solid carbon dioxide (at -79°C)
- Low-temperature deep freezers (at -80°C)
- In the vapor phase nitrogen (at -150°C)
- In liquid nitrogen (at -196°C)

Precautions

- Ice crystal formation within cells should be avoided since it causes organelle damage.
- High intracellular solute concentrations can potentially harm cells.
- Certain solutes from the cell may occasionally leak out during freezing.
- Cell viability is also affected by cryoprotectants.
- The physiological condition of the plant material is also critical.

Stage of Cryopreservation for Plant Regeneration

Cryopreservation of plant cell culture followed by plant regeneration entails the following stages:

- **Sterile tissue culture development**: increases the ability of the explant to survive cryopreservation. Cryopreservation can be done with any plant tissue. Meristems, embryos, endosperms, ovules, seeds, cultivated plant cells, protoplasts, calluses, and so on are examples.
- **Cryoprotectant addition and pretreatment**: Cryoprotectants are chemicals that can prevent cell damage caused by freezing or thawing. Dimethyl sulfoxide (DMSO), glycerol, ethylene, propylene, sucrose, mannose, glucose, proline, and acetamide are some cryoprotectants. DMSO, sucrose, and glycerol are the most commonly utilized.
- **Freezing**: The susceptibility of cells to cold temperatures differs based on the plant type.
    - **Slow-freezing method:** Plant material is slowly frozen at 0.5-5°C/min and then transferred to liquid nitrogen.

  - **Rapid freezing method:** Rapid freezing causes a temperature drop of -300° to -1000°C/min.
  - **Stepwise freezing method:** This is a combination of slow and quick-freezing procedures that are both performed in a stepwise manner.
  - **Dry freezing method:** According to certain researchers, non-germinated dry seeds can survive freezing at very low temperatures.
- **Storage:** The frozen cells and tissues are kept at a temperature between -70 and -196°C for storage.
- **Thawing:** Thawing usually happens by submerging frozen samples in calibrated quantities in a warm water bath (37-45°C) with vigorous whirling. It shields cells from the harmful effects of ice crystal production.
- **Re-culture:** Germplasm is washed multiple times after freezing to remove cryoprotectants. Following regular protocol, this material is re-cultured in a new medium.
- **Survival/viability testing:** The viability/survival of frozen cells can be tested at any stage of cryopreservation, as well as after thawing or re-culture. Triphenyl tetrazolium chloride (TTC), Evan's blue, and fluorescein diacetate (FDA) staining procedures are often utilized.
- **Plant regeneration**: Germplasm cryopreservation aims to regenerate the desired plant. Plant regeneration is frequently dependent on the availability of certain growth-promoting compounds, as well as the preservation of appropriate environmental conditions.

Applications

- Maintenance of stock culture.
- Long term conservation.
- Disease (pathogen)-free plant materials are obtained.
- Maintained for long time.
- Pollen conserve for longevity.
- Cryopreservation is an effective way for selecting cold resistant mutant cell lines that may evolve into frost-resistant plants.
- Germplasm bank establishment.

Vitrification (Hyperhydricity)

- Vitrification is an ultra-rapid freezing technique of cryopreservation that conservation of cellular water into nanocrystal- like solids which later cool rapidly.
- it is a physiological disturbance affecting tissue culture-based propagation of several plant species.

- It has glassy appearance due to absence of ice formation.

Hyperhydricity causes in plants:

- More relative humidity inside the vessels.
- Accumulation of Ethylene plant hormone.
- Due to Concentration of the gelling agents.
- Media composition (concentration of cytokinin; and other nutrients)

Major difference between cryopreservation and vitrification:

S. No.	*Cryopreservation*	*Vitrification*
1.	Low level of cryoprotectants	High level of cryoprotectants
2.	Crystal formation	No crystal formation
3.	Slow freezing	Rapid freezing
4.	Less cryoprotectants	More cryoprotectants
5.	Dehydration absent	Dehydration present
6.	Cell survival low	Cell survival high
7.	Damage due to ice formation	No damage and No ice formation
8.	Simple procedure	Complex procedure

Important Questions:

1. Define the Conservation process.
2. Describe the I situ and Ex situ conservation and its limitations.
3. What is vitrification process?
4. What is difference between cryopreservation and vitrification?

Recombinant DNA Technology

Recombinant DNA technology is a laboratory technique in which the isolation of gene of interest and gene manipulate through various enzymes. This method can be used to making DNA from different species with new creative genes functions.

Tools of recombinant DNA technology: There are some tools which are involve in making of recombinant DNA

5. **Restriction enzyme and ligase**
6. **Vector**
7. **Host organism**

1. Restriction Enzyme: Restriction enzyme is a naturally endonuclease found in prokaryotic cell which can cleave ds DNA at the specific nucleotide sequence and create the sticky and blunt site. This is defense mechanism of the bacterial and archaea cell which are protect by methylation from foreign molecule like bacteriophage.

First evidence of restriction enzyme in E.coli was discovered by Warner Arber 1962

First restriction enzyme was Hind II discovered by Hamilton O. Smith, Thomas Kelly and Kent Wilcox from *Hemophilus influenza.*

Types of Restriction Enzyme:

Title	*Sub unit*	*Cleavage site*	*Cofactor*	*Function*
Type I	Three different	Up to 1000 away from restriction site	ATP, Mg^{2+} and S-adenosyl Methionine	Bifunctional (endonuclease and methylation)
Type II	Two identical	At or near restriction site	Mg^{2+}	Unifunctional (endonuclease OR methylation)
Type III	Three different	24 to 26 bp downstream from recognize site	ATP and Mg^{2+}	Bifunctional (endonuclease and methylation)

Properties

- Restriction enzymes cleave in the form of sticky or blunt.
- Type II restriction enzyme mostly used.
- Restriction enzyme recognize GC site.
- Restriction enzyme function at 7.4 pH and 37°C.
- Palindromic sequence recognized by restriction enzyme.
- ***Eco R1*** and ***Bam H1*** produced sticky end and PvuII and AluI produced Blunt end.
- There are isoform of restriction enzymes on the basis of recognition site.
    - **Isochizomers**: Recognize same site and cut produced same or different site.
    - **Neoschizomers**: Recognized same site and cut produced different site.
    - **Isocaudomers**: Recognized different site and cut produced same site.

2. Vectors: A vector is a circular autonomous replicating DNA molecule that acts as a vehicle to transport a specific DNA segment into a host for the purpose of cloning.

Most Commonly Used Vectors

PBR 322: Plasmid Stand for Plasmid Bolivar Rodriguez. It is a first artificial cloning vector and their size of is 4.3 kb. Its Capacity of PBR 322 is 10 to 12 kb.

Phasmid: it is a DNA-based cloning vector which are formed by combining of M13 genome with plasmid DNA. These vectors carry to origin of replication derived from bacteriophage M13.

Cosmid: it is a hybrid form between a phase DNA and a bacterial plasmid

BAC (Bacterial Artificial Chromosome) Vector: It is an engineered DNA molecule which is used for DNA cloning into the bacterial cells Example E. coli. In which large piece of synthetic DNA which can be engineered that allows replicated in bacteria as a circular artificial chromosome. Each BAC consists of a DNA clone that ranges in size from 100 to 300 base pairs.

YAC (Yeast Artificial Chromosome) Vector: it is genetically engineered chromosomes that are derived from the DNA of the yeast. It is used for cloning by inserting a large fragment of DNA (1000 to 2000 kb) into yeast cell.

Vectors and their Capacity

S.No.	*Name of Vector*	*Capacity*
1.	Plasmid based vector	8 to 10 kb
2.	M13 vector	3 kb

S.No.	*Name of Vector*	*Capacity*
3.	Phasmid	10 to 12 kb
4.	Cosmid	40 kb
5.	BAC	300 kb
6.	YAC	1400 to 2000 kb

3. Host organism: Host organism are those living organism in which vector insert own DNA molecule by recombinant DNA technology through the gene transfer methods. Gene of interest carries by vector and finally insert into the host organism. In the case of plants, Host organism will be plant for the development of Transgenic (for example: - Bt Cotton, Golden Rice, GM mustard etc.).

Process of Recombinant DNA technology:

Step 1: Isolation of gene of interest

Isolate desired gene in the pure form

Step 3: Amplifying through PCR

Amplifying single copy of desired gene into thousand copies for further use. Then gene of interest has been cut using restriction enzyme.

Step 2: Gene construct

Design the gene construct with promoter, terminator, enhancer linker and adapter

Step 4: Gene insert into vector and ligation:

Gene construct insert into vector and joined by ligase enzyme

Step 5: vector insert into host

Finally recombinant vector insert into the recipient host cell called as transformation. It may be two way 1. Indirect transformation (Agrobacterium tumefaciens) 2. Direct transformation

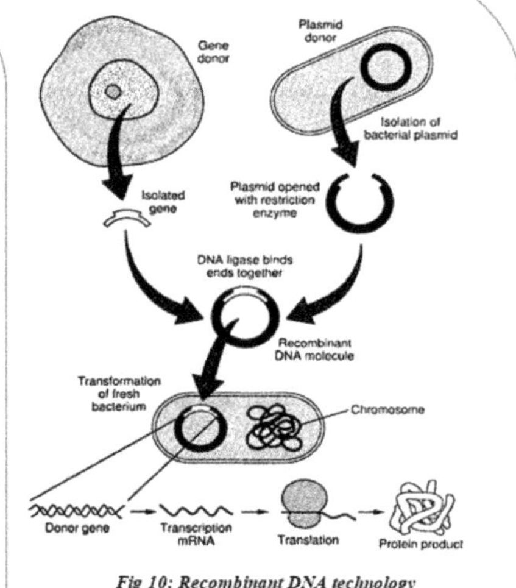

Fig 10: Recombinant DNA technology

Important Questions

1. What are Recombinant DNA Technology and its relative tools.
2. Define the pBR 322 vector.
3. Define the vectors and their capacity.
4. Difference between Isochizomers, Neoschizomers and Isocaudomers.

15 Gene Transfer Method

In the case of a plant, the Gene of interest (desired gene) can be transferred by Direct and indirect methods for the making of transgenic plant.

Method of plant transformation:

A. Indirect (Vector-mediated gene transfer)		
S. No.	**Method**	**Features**
1.	Agrobacterium-mediated gene transfer	Efficient to wide range of plants.
2.	Plant virus vectors	Efficient and high expression of transgenes.
B. Direct gene method (Physical Method)		
3.	Electroporation	Confined to protoplasts that can be regenerated to produce complete and viable plants.
4.	Microinjection	Requires highly skillful technical personnel and limited to one cell per microinjection.
5.	Particle bombardment/ microprojectile	Special instrumentation required. High risk of gene rearrangement.
6.	Silicon carbide fibers	Requires careful handling. Requires regenerable cell suspensions.
Chemical Method		
7.	Polyethylene glycol (PEG)-mediated	Confined to protoplasts. Problems encountered when regenerating these cells into viable plants.
8.	Liposome fusion	Confined to protoplasts which may be regenerated into a viable plant.
9.	Diethylaminoethyl (DEAE) dextran mediated	Does not result in stable transformation.

Important Questions

1. What is gene transfer method and their features.

16 Transgenic Crops and their Improvements

Transgenic plants are those in which the genome has been altered by genetic engineering to include one or more genes from a different species.

Purpose:

- The main purpose of the transgenic plants for Increase the ability of plants against to the stress. like disease through bacterial or fungus or so many others.
- Add a new characteristic to the plant that does not exist naturally in the species.
- A transgenic plant has a gene or genes that have been added artificially. The transgene is the introduced gene sequence; it could be from an unrelated plant or a completely different species.
- Quality and quantity improvements.
- Strong Resistance power developments.
- Less use pesticide and insecticides.
- High nutrition value.
- Increase supply of food with reduced cost and longer self-life. Like flavor sever tomato (first commercial transgenic food crop 1994)

List of Transgenic crops:

S. No.	*Year*	*Transgenic Crop*	*Name of Country*
1.	1982	Tobacco (antibiotic resistant)	China
2.	1994	flavr savr tomato	Monsanto, US
3.	1995	Bt Potato	Monsanto, USA
4.	2000	GM Rice	China
5.	2000	Golden Rice	Philippines

S. No.	Year	Transgenic Crop	Name of Country
6.	2002	Bt cotton	GEAC, India
7.	2005	Golden rice-2	Syngenta, philippines
8.	2022	GM Mustard (DHM-11)	Delhi University, India

Important Questions

1. What is transgenic plant and its purpose .
2. Write to the list of transgenic crops.

17 Molecular Markers

Markers: it associated with the particular gene/QTL traits which follow the inheritance pattern through the parents to progeny. It based on the three categories.

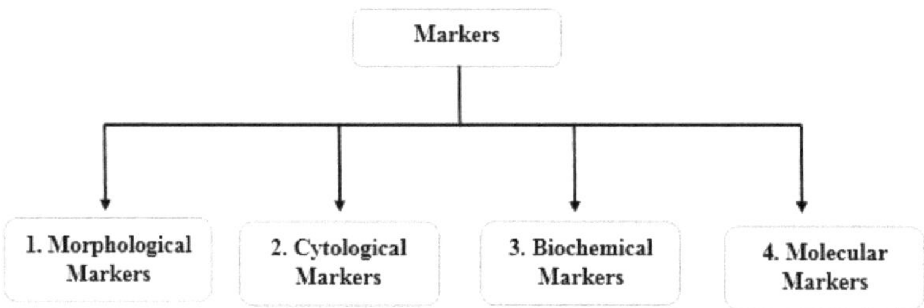

1. Morphological Markers: Visual indicators of phenotypically distinct characteristics such as the colour, shape, and size of the flower, seeds, or leaves are known as morphological markers.

2. Cytological Markers: Cytological markers are chromosomal bands produced by different stains, such as G banding, and they are connected to changes in chromosome number, shape, size, and banding pattern.

3. Biochemical Markers: Biochemical markers is peptide polymorphism that are produced by the various gene. it is used for the detect to polymorphic changes at the amino acid level. Example: Alloenzymes, isozyme, Amylase, plasma protein, ALT (alanine transaminase), AST (Aspartate aminotransferase), alfa fetoprotein etc.

4. Molecular Markers: Molecular markers are small pieces of DNA that are associated with a certain position within the genome and are used to identify a specific nucleotide sequence of DNA in a pool of unknown DNA.

Codominant and Dominant Molecular Markers

Co-dominant Markers: Codominant markers are those for which both alleles (AA), (Aa) and (aa) are expressed in a F1 Progeny. Its separation is enabled in the form of allele and genotype determination.

Dominant Markers: in this case, identify single dominant allele (AA), (Aa) in F1 progeny.

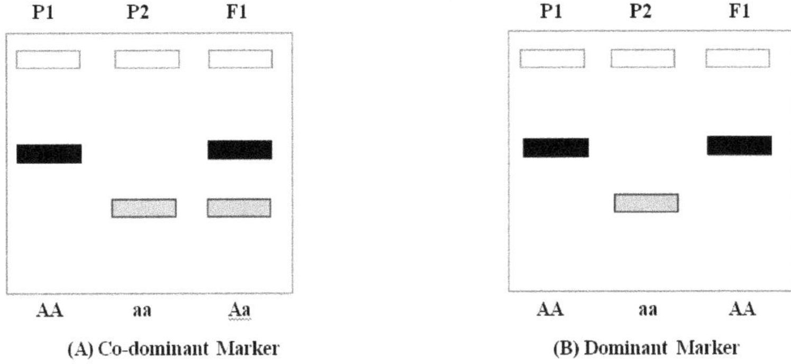

Fig. Identify Co-dominat and dominat marker in gel electrophoresis

Markers and their Features:

S. No.	*Markers*	*Features*
1.	RFLP	• It is hybridized based co-dominant marker • Identified the variation through the polymorphism within specific nucleotide sequence generated by specific endonuclease from specific locus • No sequence information
2.	AFLP	• Dominant marker and PCR based • Detection by variable length of DNA using restriction enzyme
3.	RAPD	• Dominant marker and PCR based • Polymorphism generated using random primer
4.	SCAR	• Co-dominant and PCR based • Amplification by PCR using specific primer
5.	SSR	• Co-dominant and PCR based • Used for genetic diversity
6.	SNP	• Co-dominant • Variation in single nucleotide

Difference between Molecular Markers:

Marker/ Features	*PCR based*	*Dominancy*	*Reproducibility*	*Analysis cost*	*DNA Quality*
RFLP	No	Co-dominant	High	High	High
AFLP	Yes	Dominant	High	Moderate	Moderate
RFLP	Yes	Dominant	High	Low	High
SSR	Yes	Co-dominant	High	Low	Moderate
SNP	Yes	Co-dominant	High	Low	High

Full Form:

RFLP	Restriction Fragment Length Polymorphism
AFLP	Amplified Fragment Length Polymerase
RAPD	Random Amplified Polymorphism DNA
SCAR	Sequence Characterized Amplified Region
CAPS	Cleaved Amplified Polymorphism Sequence
VNTR	Variable Number Tandem Repeats
SSR	Simple Sequence Repeats
SNP	Single Nucleotide Polymorphism
ISSR	Inter Simple Sequence Repeats

Important Questions:

1. What are molucular markers and how they are useful for crop improvement explain with example.

18 Polymerase Chain Reaction (PCR)

Polymerase chain reaction is a laboratory technique to make many copies of a specific DNA fragments. It is based on the ability of DNA polymerase to create new DNA strands that are complementary to the provided from template strand.

Requirements for PCR: There are some components that are involved in this process.

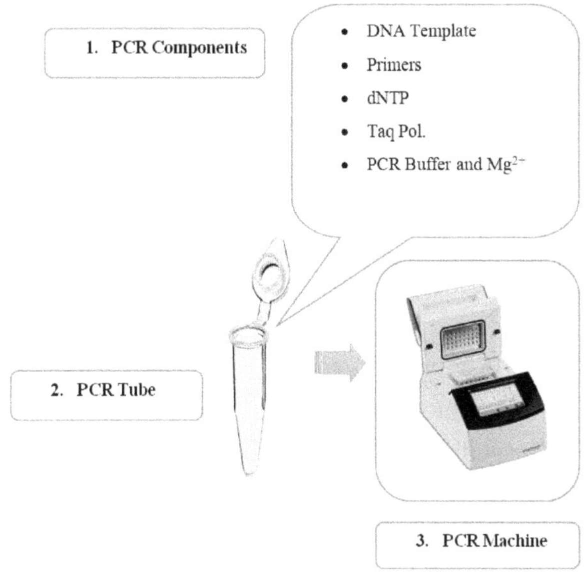

Features of PCR Components:

- **DNA templates:** Desired DNA template should be appropriate for PCR reaction up to 3 kb
- **Primer:**
  - Primer should be forward and reverse
  - It should be 18 to 24 nucleotides

  - GC content of primer should be 45-55%
  - Melting temperature (Tm) = [4 x (G+C) + 2 x (A+T)]
  - Annealing temperature below from melting temperature
- **dNTP:** di nucleotide tri phosphate is charge nucleotide base A, T, G, C required for synthesis of new DNA strand.
- Taq Polymerase:
  - It is isolated from thermostable bacteria (*Thermophilus aquatic*)
  - It is mostly used for DNA amplification in PCR
  - Taq pol. bind at the 3' OH site but lack the 3' to 5' exonuclease activity
- **PCR Buffer and $MgCl_2$:** PCR buffer and magnesium ion are required for balancing

Steps of PCR reaction: There are three major steps

1. Denaturation (94 °C for 1 Minute):

During this process, denature of double stand DNA separate in two single strand form

2. Annealing (54 °C for 45 Second):

Primer are attached to both single stranded template through ionic bonding at the 3'OH after than Taq pol. are attached and ready for polymerization of nucleotide.

3. Primer Extension (72 °C for 2 Minute):

During this process, Addition of nucleotide with help of Toq pol. formed two double strand of DNA in one cycle.

Application of PCR:

- Amplification of gene
- Determination of physical location of gene
- Study of transgenic and DNA fingerprinting

Important Questions:

1. What is PCR process and their relative steps and componants?
2. What are applications of PCR?

19 Marker Assisted Breeding

- **Genome mapping**
- **QTL mapping**
- **Marker-assisted Selection**

1. Genome Mapping:

Identifying the position of a certain gene on a chromosome and the relative distances between genes on the chromosome. Two different types of genomic maps exist.

Linkage maps: A linkage map, also known as a genetic map, depicts the relative positions of genes on a chromosome. It is based on the genetic linkage principle, which states that genes that are physically close to each other on a chromosome are more likely to be inherited together than genes that are farther apart.

The distance between genes in a linkage map is quantified in map units (centimorgans or cM). One map unit indicates a 1% recombination frequency, hence genes with a 1% recombination frequency are one map unit apart.

Physical maps: A physical map depicts the exact physical positions of DNA segments on a chromosome or genome. A physical map, as opposed to a linkage map, which is based on genetic linkage and recombination frequencies, offers exact information about the physical distances between DNA markers or genes, which are often quantified in base pairs (bp) or kilobases (kb).

There are two types of physical maps

- **Contig Maps**: Contig maps show the arrangement of overlapping DNA fragments or clones that cover a specific region of a chromosome or genome. These maps are often created using techniques such as DNA sequencing, restriction mapping, and clone fingerprinting.
- **Sequence Maps**: Sequence maps are based on the actual DNA sequences of a chromosome or genome. With advances in DNA sequencing technologies, it has become possible to obtain the complete nucleotide sequence of entire genomes.

2. Qualitative Trait Loci (QTL)

- Qualitative Trait Loci (QTL) analysis is a crucial process in genetic studies aimed at identifying genetic regions associated with qualitative traits in organisms.
- Gelderman (1919) invented the term QTL (Quantitative Trait Locus), which is described as "a region of the genome or locus of gene that is associated with an effect on a quantitative trait." Plant height, crop output, colour, seed size, and so on.
- A quantitative trait that is controlled by numerous genes, each of which has a minor effect, is additive in nature, and is influenced by the environment. A bell curve is commonly used to show the phenotypes of quantitative traits.

Rules for Choosing the Best Targeted QTL

- **Size of effect**: In order to prevent linkage drags, which result in the selection of undesired qualities together with the gene for a given trait, the distance between the marker and the gene of interest should be extremely small. Rules for choosing the best targeted QTL
- **Interactions with the environment**: The majority of genes that have been discovered as quantitative traits are heavily influenced by environments that are polygenic in character and exhibit cumulative behavior in a variety of contexts.
- **Environmental interactions**: The majority of genes identified as quantitative traits are extensively influenced by environments and exhibit cumulative behavior under various environmental conditions.

QTL Mapping

- The method of locating genes with impacts on quantitative traits using molecular markers is known as QTL mapping.
- QTL mapping is a statistical approach that finds the molecular markers that result in a quantitative change of a certain trait.

Principle

The principle of QTL mapping involves several key steps:

- **Trait Measurement**: The first step is to measure the quantitative trait of interest in a population or a group of individuals. Traits can be anything that varies quantitatively, from physical characteristics to complex phenotypes like disease susceptibility or yield in crops.
- **Mapping Population**: A mapping population is created by crossing two genetically distinct parents, often referred to as the "founder" or "parental" lines. This results in a genetically diverse group of offspring, each carrying a unique combination of the parental genomes.

- **Genotyping**: The individuals in the mapping population are genotyped at multiple genetic markers spread throughout the genome.
- **Phenotype-Genotype Association**: Statistical analysis is performed to identify correlations between the genotypes at specific genetic markers and the observed variation in the quantitative trait.
- **Linkage Analysis**: QTL mapping relies on the principle of genetic linkage, which describes the tendency of genetic markers and nearby genes to be inherited together due to their physical proximity on the same chromosome.
- **QTL Identification**: By analyzing the patterns of trait variation and genetic marker inheritance across the mapping population, researchers can identify regions of the genome (QTLs) that are statistically associated with the trait.
- **QTL Validation**: Once potential QTLs are identified, further experiments and validation studies are conducted to confirm the association between the QTLs and the quantitative trait. This can involve additional genetic crosses or manipulation of the candidate genes to determine their effects on the trait.

Requirements

- The major requirements of QTL are Mapping Population, Linkage Map, Phenotypic Screening and Statistical analysis.
- The Marker should be less than 5 centimorgans from a gene or QTL.
- QTL mapping procedures include selecting different parents. Creating mapping populations for producing saturated linkage maps, phenotype screening, and contrasting the mean of the marker attached and pair of gene of interest at each marker locus.
- The primary goal of this study is to provide a straightforward method for investigating the number of genes that influence the trait. The location of the gene that affects traits of interest, in order to determine the effect of genes on trait variation.

3. Marker Assisted Selection (MAS)

Marker-Assisted Selection (MAS) is a powerful breeding technique that combines traditional breeding methods with modern molecular genetics. It involves the use of molecular markers linked to desired traits to expedite the breeding process, leading to the development of improved and desirable plant or animal varieties.

QTL Mapping Significance for MAS

The importance of QTL mapping for MAS can be summarized as follows:

- **Identifying Genes and Loci Linked to Desired Traits**: QTL mapping allows researchers to identify specific genomic regions (QTLs) associated with the expression of target traits.
- **Enhancing Precision in Trait Selection**: QTL mapping helps in prioritizing and selecting the most promising individuals or parents based on their genetic makeup for targeted breeding.
- **Improving Breeding Efficiency**: The uses molecular markers closely linked to the genes controlling the traits of interest. This significantly speeds up the selection process indirectly select for specific genes without the need to evaluate the actual trait expression in each individual.
- **Enhancing Genetic Gain**: By integrating QTL mapping into MAS, breeders can focus on specific genomic regions associated with high heritability and significant trait variation.
- **Supporting Genomic Selection**: QTL mapping is a fundamental step in building the reference genomic map required for genomic selection.

MAS in Plant Breeding:

Marker Assisted Backcrossing: There are three major levels of selection process

- **1st Lavel**: In this level, Screening of the target trait by molecular markers
- **2nd Level**: This level is a recombinant selection level in which progeny are selected with target gene
- **3rd Level**: In this level, progeny are selecting by Backcross with background marker

Marker Assisted Pyramiding:

Pyramiding is the process of integrating many genes/QTLs into a single genotype at the same time. This is achievable by traditional breeding, however at the early generations it is extremely difficult or impossible.

Difference between Conventional Breeding and Marker Assisted Selection

S. No.	*Conventional Breeding*	*Marker Assisted Selection*
1.	Phenotype selection based on breeding process	Phenotype selection based on molecular marker
2.	Long time process	Short time process
3.	Selection in BC20	Selection in BC 2
4.	Traditional method	Advanced method
5.	Selection on healthy plant	Selection on DNA Banding process

Important Questions:

1. What is QTL Mapping and their principles?
2. What is MAS and its selection process?
3. Difference between conventional and MAS.

20 List of Apparatus and Equipment's

A. Laboratory Apparatus

Commonly Used Laboratory Apparatus

1. Beakers	2. Conical flasks
3. Boiling flasks	4. Test tubes, Tongs and Racks
5. Watch Glasses	6. Crucibles
7. Funnels	8. Graduated Cylinders
9. Volumetric Flasks	10. Droppers
11. Micro Pipettes	12. Burette
13. Ring Stands, Rings and Clamps	14. Forceps
15. Spatulas	16. Thermometers
17. Bunsen Burners	18. Balance

1. Beakers

- It is used for mixing, stirring, and heating chemicals.
- They typically have lips around their rims and marks to estimate the volume they contain, though these are not precise

2. Conical Flasks

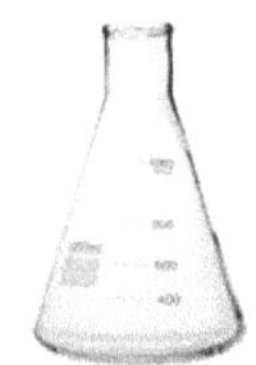

- The thin neck and enlarged base enable for easy mixing with minimum spilling; the tight opening allows for the use of a stopper
- They can be simply clamped to ring stands, heated, or mechanically shook

3. Boiling Flasks

It has a spherical bottom and a long neck, and it can be readily swirled and heated

- It can also be easily capped by rubber or glass stoppers

4. Test Tubes with Stand

- Test Tubes are used for the plant tissue culture and fungus isolation. It is a glass tube with one end open only that occurs upper side
- They are usually stored on a test tube rack
- If the test tubes get too hot to handle, use a test tube tong to shift them.

5. Watch Glasses

- A spherical, slightly concave/convex piece of glass that may retain a little amount of liquid or solid
- It can be used for evaporation and also as a cap for a beaker

6. Crucibles

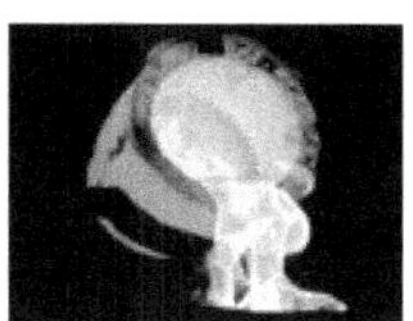

- It is a little clay cup constructed of a heat-resistant material that is used for heating substances and comes with lids.

7. Funnels and Filter Paper

- It is a laboratory funnel used to separate particulates from liquids using the laboratory filtering technique.

8. Graduate Cylinders

- Its used for measuring volume of a liquid
- It is different type based on length of container with specific increments

9. Volumetric Flasks

- It is a spherical flask with a long neck and a flat bottom that is used to precisely measure the volume of liquid.

10. Droppers

- These are thin glass tubes with narrow tips on one end and a rubber bulb on the other that suck up liquid and then squeeze out small drips

11. Micropipettes

Micro-pipettes are designed to pipette specified volume and are used to measure an exact volume of liquid and transfer it to another container

12. Burette

- Burette is a glass tube with an open top and a narrow-pointed hole at the bottom. It is used for exceptionally accurate liquid addition

13. Ring Stand, Ring and Clamps

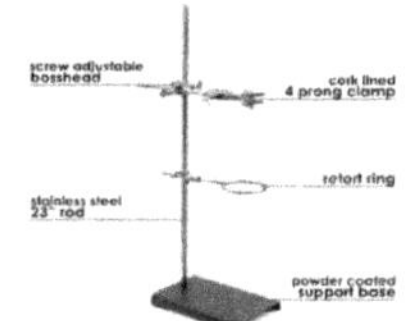

- Burettes are suspended using the ring stand.
- Beakers, flasks, crucibles, and other similar things are held in place by rings

14. Forceps

Forceps are used to grip small things, such as broken-up solid chemicals handled and added to containers securely by them

15. Spatulas

- Spatulas are used to scoop solid chemicals out of their original containers and onto a weigh boat/paper. so that it can be balanced

16. Thermometer

- Thermometer is made by glass. It is used for measuring the temperature of liquid.

17. Bunsen Burners

- Burners is a type of mechanical apparatus that is used for sterilize forceps and other apparatus during explant inoculation in laminar air flow
- It connects to flammable gas source
- It kills the all type of microorganism who create the contamination in culture

18. Balances

- Electric balance is used to measurement of chemicals. Two doors are present that closed during weigh a chemical
- When we will use them for weigh than doors should be closed

B. Laboratory Equipment's

Commonly Used Laboratory Equipment

1. Centrifuge	2. Urinometer
3. Electrophoresis Assembly	4. Colorimeter
5. Spectrophotometer	6. Fluorometer
7. Luminometer	8. Autoanalyzer
9. pH Meter	10. Thermal Cycler
11. HPLC	12. Autoclave
13. Incubators	14. Hot air Oven
15. Water bath	

1. Centrifuge

- Centrifuge is a laboratory equipment that are used for separation on the basis of density such as DNA isolation, RNA isolation etc.

2. Urinometer

- Its used to calculate the specific gravity of urine

3. Gel Electrophoresis

The migration and separation of charged particles (ions) under the influence of an electric field is referred to as electrophoresis

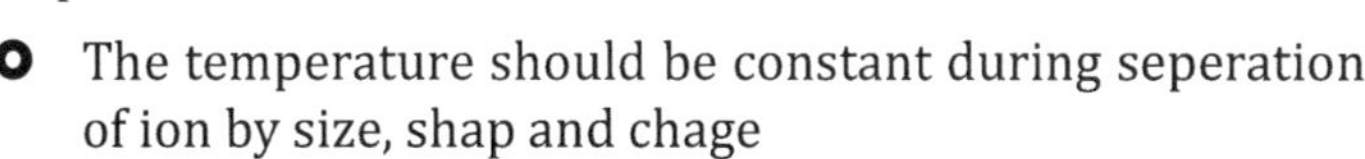

- The temperature should be constant during seperation of ion by size, shap and chage

4. Colorimeter:

- It is a device that assists specific solutions in absorbing a specific wavelength of light
- It uses the Beer-Lambert law to determine the concentration of a known solute in a given solution

5. Spectrophotometer

- A spectrophotometer is used to detect any light-absorbing particles dissolved in a solution
- The spectrophotometer directs a whole spectrum of white light towards a container containing a liquid sample

6. Florescence

- It is a type of spectrometer
- In this equipment's, when a sample is excited (irradiated) with one wavelength of light and emits another, usually longer wavelength of light
- A fluorometer is also known as a fluorimeter

7. Luminometer

- It monitors the amount of light produced during particular chemical reactions.

8. Autoanalyzer

- Autoanalyzer is a laboratory equipment that is used for measures various constituents of the body with minimal human assistance.

9. pH Meter

- It detects the concentration of hydrogen ions in a solution, indicating its alkalinity or acidity, which is denoted as pH.

10. Thermal Cycler

- It is a PCR machine that is used for amplify gene sequence or DNA fragments
- PCR is also used as c DNA library, Genomic library etc.

11. HPLC

- HPLC stand for High Performance Liquid Chromatography
- HPLC is a powerful analytical technique used to separate, identify, and quantify components in a mixture.

12. Autoclave

- Autoclave is the most important equipment of biotechnological lab which is used for all the microorganism kill at high steam pressure.
- Any apparatus should be sterilized before using
- High steam pressure of autoclave should be required 121°C at 15 psi (Pressure per square inch) for 15 to 20 minutes.
- Autoclave is mostly of two types: 1. Horizontal 2. Vertical

13. Incubator

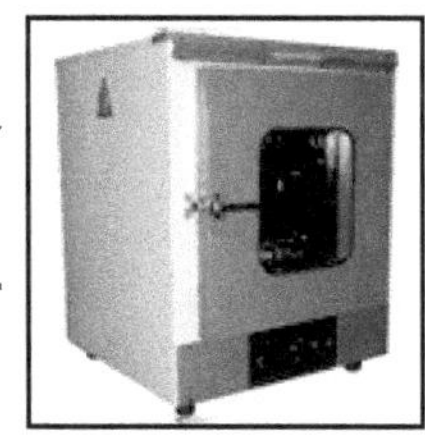

- It is a laboratory equipment which is used for grow and maintain the microbial cell culture
- It provides to optimum temperature and humidity for the microbial growth

14. Hot air oven

- It is extremely high temperatures equipment of the biotechnology lab which is used for sterilize items by heating the outside surfaces. And destroy microorganisms and bacterial spores also.

15. Water Bath

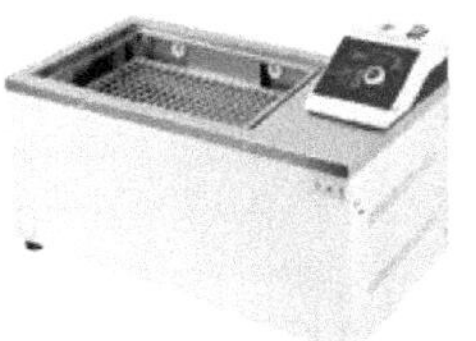

- It is a piece of laboratory equipment consisting of a container filled with hot water that is used to incubate samples in water at a steady temperature for an extended period of time.
- Applications include reagent warming, substrate melting, and cell culture incubation. It's also utilized to allow some chemical reactions to take place at high temperatures.

Multiple Choice Questions

Part -1

1. **The most common solidify agent used for the micropropagation is**
   (a) Dextran (b) Agar
   (c) Mannan (d) All of these
2. **Which of the following the most common method for the development of virus free plants**
   (a) Meristem culture (b) Embryo culture
   (c) Anther culture (d) All of these
3. **Liquid agitated media for cell culture is**
   (a) Suspension culture (b) Micro propagation
   (c) Agar culture (d) Anther culture
4. **Batch culture is , where**
   (a) Medium is loaded only at beginning
   (b) Medium is continuously
   (c) Occurs sigmoid curve
   (d) Both (a) and (c)
5. **All are plant derived except**
   (a) Codeine (b) Quinine
   (c) Nicotine (d) Menthol
6. **All are plant derived elicitor except**
   (a) Pectic acid (b) Cellulose
   (c) Pectin (d) Chitin

7. **Artificial seed are**
   (a) Somatic embryo encapsulated in a gel
   (b) Zygotic embryo encapsulated in a gel
   (c) Seed encapsulated in a gel
   (d) All of these
8. **Crown gall disease transfer through**
   (a) Agrobacterium tumefaciens
   (b) Agrobacterium rhizogenes
   (c) Virus
   (d) E. coli
9. **Variation in *in-vitro* culture**
   (a) Somaclonal variation (b) Mutation
   (c) Haploid culture (d) All of these
10. **Somaclonal variation is detected by**
    (a) Genetic fidelity (b) Gene diversity
    (c) Both (a) and (b) (d) None of these
11. **Hybrid developed under plant tissue culture through**
    (a) Fusion of two cytoplasm
    (b) Fusion of two nucleus only
    (c) Fusion of one protoplast and enucleated protoplast
    (d) None of these
12. **Fusion of one protoplast with enucleated protoplast of other plant is known as**
    (a) Cybrid (b) Hybrid
    (c) Haploid (d) All of these
13. **Organogenesis is**
    (a) Root and Shoot formation (b) Callus formation
    (c) Totipotency (d) All of these
14. **Recombinant DNA technology is**
    (a) Fusion of one gene insert into other organism
    (b) Gene construct
    (c) Design of vector
    (d) All

15. Protoplast fusion by

(a) PEG (b) G3P

(c) Ribulose (d) Pectinase

16. Asexual reproduction is also called

(a) Clone (b) Assembly

(c) Meiosis (d) All of these

17. Restriction enzyme is used for

(a) Cut produced double standard at specific sequence

(b) Cut produced single standard only

(c) Sticky end only

(d) Only blunt end

18. Restriction enzyme is also called

(a) Molecular glue (b) Molecular scissors

(c) Both (d) None of these

19. DNA ligase is also called

(a) Molecular or genetic glue (b) Molecular scissors

(c) DNA gyrase (d) Topoisomerase

20. DNA ligase is

(a) Joint the two fragments (b) Cut produced

(c) Break hydrogen bond (d) None of these

21. First plasmid used for the production of recombinant insulin is

(a) PBR 322 (b) Ti plasmid

(c) pUC 18 (d) Phasmid

22. BT Cotton is a

(a) Transgenic crop (b) Wild crop

(c) Both (a) and (b) (d) None of these

23. A device in which a substrate of low value is utilized by living cells or enzymes to generate a product of higher value is called

(a) Test tube culture (b) Bioreactor

(c) Autoclave (d) Electrophoresis

24. Bioreactor is used for

(a) Low to high production (b) High to low production

(c) Kill microorganism (d) All of these

25. Agar is a

(a) Solidifying agent (b) Antibiotic

(c) Growth factor (d) Liquid form

26. Explant is

(a) Any cut part of plants (b) Plant growth regulator

(c) Plant extract (d) All of these

27. Totipotency is

(a) Ability of a plant cell produced whole plant

(b) Development of embryo

(c) Heart stage of embryo

(d) All of these

28. Somatic embryo are

(a) Embryos developed from zygote after fertilization

(b) Embryos developed from egg without fertilization

(c) Embryo like structure developed from the cells of callus

(d) Embryo developed by ovules

29. An amorphous mass of loosely arranged thin-walled parenchyma cells developing from explant is called

(a) Callus (b) Explant

(c) Embryo sac (d) Seed

30. Meristem culture is used for

(a) Virus free plants (b) Hybrid plant

(c) Cybrid (d) all of these

31. Genetic variation observed by

(a) Somaclonal ariation

(b) Cell suspension culture

(c) Secondary metabolite production

(d) all of these

32. Secondary metabolite production by

(a) Cell suspension culture (b) Somaclonal variation

(c) Both (a) and (b) (d) None of these

33. Vitamin A is a richest source of

(a) Golden rice (b) Banana

(c) Litchi (d) Grapes

34. Father of plant tissue culture

(a) G. haberlandt (b) Went

(c) White (d) Mendal

35. First plant hormone is

(a) Auxin (b) Ethylene

(c) Abscisic acid (d) Gibberellic acid

36. Auxin is used for

(a) Root formation (b) Shoot formation

(c) Callus (d) All of these

37. Size of explant

(a) 1.0 to 1.5 cm (b) 2.0 to 3.0 cm

(c) 3.0 to 4.0 (d) 4.0 to 5.0

38. First plant tissue culture media are

(a) MS Media (b) LS Media

(c) B_5 Media (d) White Media

39. Agar is obtained from

(a) White algae (b) Red algae

(c) Blue algae (d) None of these

40. Natural amino acid is

(a) Glycine (b) Arginine

(c) Aspartic acid (d) Cysteine

41. Callus produced to the whole plants

(a) Redifferentiation (b) Dedifferentiation

(c) Regeneration (d) Adventitious

42. Undifferentiated mass of cell

(a) Explant (b) Callus

(c) Both (d) None of these

43. Dedifferentiation is

(a) Explant produced to callus (b) Callus produced to embryo

(c) callus produced to plant (d) All of these

44. Shoot or Roots developed from unusual points

(a) Adventitious (b) Regeneration

(c) Totipotency (d) None of these

45. Embryo rescue is

(a) Isolate embryo before deadline condition of plants

(b) Protects to the embryo

(c) Embryo safe for next generation

(d) All of these

46. How many concentrations of ethanol is used for the surface sterilization

(a) 70% (b) 80%

(c) 90% (d) 100%

47. Autoclave is sterilized at

(a) 121°C at 15 psi (Pound per square inch) for 20 minutes

(b) 121°C at 15 psi (Pound per square inch) for 10 minutes

(c) 121°C at 10 psi (Pound per square inch) for 40 minutes

(d) 121°C at 5 psi (Pound per square inch) for 20 minutes

48. Teepol is used for

(a) Washing to apparatus (b) Dry to apparatus

(c) Hot to apparatus (d) None of these

49. Triploid is

(a) 2n (b) 3n

(c) 5n (d) 6n

50. Seedless fruit is

(a) Diploid (b) Triploid

(c) Tetraploid (d) Hexaploidy

[Part 1 – Answer Key]

Q.	*Ans.*	*Q.*	*Ans.*	*Q.*	*Ans.*	*Q.*	*Ans.*
1.	(b)	**14.**	(d)	**27.**	(a)	**40.**	(a)
2.	(a)	**15.**	(a)	**28.**	(c)	**41.**	(a)
3.	(a)	**16.**	(a)	**29.**	(a)	**42.**	(b)
4.	(d)	**17.**	(a)	**30.**	(a)	**43.**	(a)
5.	(d)	**18.**	(b)	**31.**	(a)	**44.**	(a)
6.	(d)	**19.**	(a)	**32.**	(a)	**45.**	(d)
7.	(a)	**20.**	(a)	**33.**	(a)	**46.**	(a)
8.	(a)	**21.**	(a)	**34.**	(a)	**47.**	(a)
9.	(a)	**22.**	(a)	**35.**	(a)	**48.**	(a)
10.	(a)	**23.**	(b)	**36.**	(a)	**49.**	(b)
11.	(a)	**24.**	(a)	**37.**	(a)	**50.**	(b)
12.	(a)	**25.**	(a)	**38.**	(a)		
13.	(a)	**26.**	(a)	**39.**	(b)		

Multiple Choice Questions

Part- 2

1. **Clonal propagation developed by**
   (a) Morel 1960 (b) Went
   (c) Carlson (d) None of these
2. **Micropropagation is also called**
   (a) Clonal propagation (b) Vegetative propagation
   (c) Asexual propagation (d) All of these
3. **Somatic hybridization was first introduced by**
   (a) Carlson (b) Morel
   (c) Mendel (d) All of these
4. **Protoplast term given by**
   (a) Klerker (b) Carlson
   (c) Hanstein (d) None of these
5. **Androgenesis means**
   (a) Culture of anther (b) Culture of ovary
   (c) Culture of ovule (d) All of these

6 **Gynogenesis means**
   (a) Culture of ovary/ovule (b) Culture of anther or pollen
   (c) Both (d) None of these

7. **Single cell culture is also called**
   (a) Anther culture (b) Cell suspension culture
   (c) Meristem culture (d) All of these
8. **Sigmoid curve belong to which type of culture**
   (a) Anther culture (b) Apical meristem culture
   (c) Micropropagation (d) Batch culture

9. Which phase of batch culture whose belong to the increase the number of cell

(a) Lag phase (b) Log phase

(c) Stationary phase (d) Decline phase

10. Parabolic growth pattern occurs in

(a) Batch culture (b) Continuous culture

(c) Lag phase (d) None of these

11. DMSO is

(a) Permeabilizing agent (b) Gelling agent

(c) Solidifying agent (d) None of these

12. Somaclonal variation term given by

(a) Larkin and Scowkraft 1981 (b) Morel and Martin 1952

(c) Guha and Maheshwari (d) None of these

13. Cell without cell wall called

(a) Cytoplasm (b) Hyaloplasm

(c) Protoplasm (d) Protoplast

14. Optimum pH of protoplast isolation

(a) 4.5 to 6.5 (b) 5.5 to 6.0

(c) 6.0 to 7.0 (d) >7.0

15. Optimum pH of protoplast culture

(a) 4.5 to 6.5 (b) 5.5 to 6.0

(c) 6.0 to 7.0 (d) >7.0

16. Hemicellulose isolated from

(a) Aspergillus sp. (b) Trichoderma viridae

(c) Arthobacter (d) None of these

17. "Pusa jai kisan" is a variety of

(a) Somaclonal (b) Cell suspension culture

(c) Hybrid (d) Cybrid

18. Preservation of the natural resources is called as

(a) Preservation (b) Conservation

(c) Deep freeze (d) None of these

19. Preservation in the frozen state is

(a) Preservation (b) Cryopreservation

(c) Deep freeze (d) None of these

20. An ultra-rapid freezing technique of cryopreservation that conservation of cellular water into nanocrystal- like solids which later cool rapidly called as

(a) Preservation (b) Conservation

(c) Vitrification (d) None of these

21. Vitrification is also called

(a) Hyperhydricity (b) Preservation

(c) Conservation (d) None of these

22. Vitrification is a type of

(a) Slow freezing (b) Rapid freezing

(c) Slow freezing to high (d) None of these

23. No ice crystal formation occurs

(a) Vitrification (b) Cryopreservation

(c) Both (d) None of these

24. Gene isolated from one organism and insert into another organism called

(a) Gene cloning (b) Genetic engineering

(c) Hybridization (d) None of these

25. Gene formed by joining of DNA segments from two different sources are called as

(a) Gene construct (b) Recombinant gene

(c) Chimeric gene (d) None of these

26. Which of the following enzyme is used for cut to DNA molecule at specific site

(a) Ligase (b) Phosphatase

(c) Ribonuclease (d) Restriction endonuclease

27. Who discovered the restriction endonuclease

(a) Nathan, Arber & Smith 1970 (b) Watson and crick

(c) Paul berg 1975 (d) Boyer and Cohen 1975

28. First recombinant DNA molecule created by

(a) Nathan, Arber & Smith 1970 (b) Watson and crick

(c) Paul berg 1975 (d) Boyer and Cohen 1975

29. The DNA molecule to which the gene of interest is integrated for cloning called as

(a) Vector (b) Host

(c) Transformer (d) None of these

30. Which is DNA segment that is used for cloned called as

(a) DNA insert (b) Gene of interest

(c) Gene fragment (d) None of theses

31. First successful transformation of recombinant DNA molecule into bacterial host was carried by

(a) Nathan, Arber & Smith 1970 (b) Watson and crick

(c) Paul berg 1975 (d) Boyer and Cohen 1975

32. Which Plasmid used by Cohen and Boyer for own first experiment was

(a) pSC101 (b) PUC 17

(c) E. coli (d) pBR322

33. Size of PBR vector

(a) 4.0 kb (b) 4.2 kb

(c)4.3 kb (d) None of these

34. Gene cloning is

(a) Production of large number of copies

(b) Production of small number of copies

(c) Both (a) and (b)

(d) None of these

35. Recombinant DNA technology was discovered by

(a) Stanley Cohen and Herbert Boyer

(b) Nathan, Arber and Smith 1970

(c) Watson and crick

(d) Paul berg 1975

36. Virus mediated gene transfer method occurs

(a) Transduction (b) Transfection

(c) Transformation (d) None of these

37. Natural gene transfer method is

(a) Agrobacterium species (b) Agrobacterium radiobactor

(c) Thermus aquaticus (d) None of these

38. Agrobacterium tumefaciens is a

(a) Gram negative (b) Gram positive

(c) Both (d) None of these

39. Agrobacterium tumefaciens cause

(a) Crown gall disease (b) Harry root disease

(c) Both (d) None of these

40. Which gene carry to gene of interest into host plant by agrobacterium tumefaciens

(a) T-DNA (b) Vir gene

(c) Both (d) Ti plasmid

41. Ti plasmid vector include

(a) Binary vector (b) Cointegrated vector

(c) Both (a) and (b) (d) None of these

42. Which of the following chemical compound enhance to vir gene expression by plants

(a) Cyanidin (b) Dextran

(c) Acetosyringone (d) None of these

43. Introduction of DNA into cells by exposing to high voltage electric pulse is

(a) Electroporation (b) Electrolysis

(c) Both (a) and (b) (d) None of these

44. Injection of DNA into developing inflorescence using a hypodermic syringe is called

(a) Macroinjection (b) Micromanipulation

(c) Microinjection (d) None of these

45. First transgenic crop developed by

(a) India (b) China

(c) America (d) Japan

46. First transgenic plant is

(a) Cotton (b) Tobacco

(c) Maize (d) All of these

47. First transgenic commercial crop is

(a) Potato (b) Flavr savr tomato

(c) Banana (d) None of these

48. Golden rice is

(a) Biofortification (b) Disease resistance

(c) Heat tolerance (d) All of these

49. Flavr savr tomato developed from the which technique

(a) RNAi (b) FISH

(c) GISH (d) All of these

50. Golden rice is richest source of

(a) Vitamin A (b) Vitamin C

(c) Vitamin D (d) All of these

[Part 2 – Answer Key]

Q.	*Ans.*	*Q.*	*Ans.*	*Q.*	*Ans.*	*Q.*	*Ans.*
1.	(a)	**14.**	(a)	**27.**	(a)	**40.**	(b)
2.	(a)	**15.**	(b)	**28.**	(c)	**41.**	(c)
3.	(a)	**16.**	(a)	**29.**	(a)	**42.**	(c)
4.	(c)	**17.**	(a)	**30.**	(a)	**43.**	(a)
5.	(a)	**18.**	(b)	**31.**	(d)	**44.**	(a)
6.	(a)	**19.**	(b)	**32.**	(a)	**45.**	(b)
7.	(b)	**20.**	(c)	**33.**	(c)	**46.**	(b)
8.	(d)	**21.**	(a)	**34.**	(a)	**47.**	(b)
9.	(b)	**22.**	(b)	**35.**	(a)	**48.**	(a)
10.	(b)	**23.**	(a)	**36.**	(a)	**49.**	(a)
11.	(a)	**24.**	(b)	**37.**	(a)	**50.**	(a)
12.	(a)	**25.**	(c)	**38.**	(a)		
13.	(d)	**26.**	(d)	**39.**	(a)		

Multiple Choice Questions

Part- 3

1. **RFLP is a**
   (a) Dominant marker (b) Codominant and hybridized based marker
   (c) PCR based (d) None of these
2. **AFLP is a**
   (a) Dominant marker (b) Codominant marker
   (c) PCR based (d) None of these
3. **SSR is a**
   (a) Dominant marker (b) Codominant marker
   (c) PCR based (d) None of these
4. **SSR is mostly used for**
   (a) Genetic diversity (b) Gene identification
   (c) Hybridized marker (d) None of these
5. **SNP is a**
   (a) Dominant marker (b) Codominant marker
   (c) PCR based (d) None of these
6. **Hybridized based molecular marker is**
   (a) AFLP (b) RFLP
   (c) SSR (d) RAPD
7. **RFLP, SSR, SNP, SCAR and CAPS are**
   (a) Dominant marker (b) Codominant marker
   (c) PCR based (d) None of these
8. **What is Molecular Marker**
   (a) Fragment of DNA (b) Fragment of protein
   (c) Both (a) and (b) (c) None of these

9. Molecular marker is used for

(a) Physical maps (b) Cytogenetic maps

(c) Chromosome maps (d) All of these

10. RFLP is used to

(a) Construct high resolution linkage maps

(b) Identified single gene disease

(c) Construct QTL maps

(d) all of these

11. RFLP is used for

(a) Specific protein (b) Specific DNA

(c) Specific RNA (d) None of these

12. PCR was discovered by

(a) Kary Mullis (b) Altman

(c) Milstein (d) None of these

13. PCR reaction is

(a) DNA sequencing technique (b) DNA amplification technique

(c) DNA degradation technique (d) All of these

14. Taq pol. is isolated from

(a) Agrobacterium sp. (b) Thermus aquatics

(c) Both (d) None of these

15. How many steps are involved in PCR

(a) 1 (b) 2

(c) 3 (d) 4

16. First step of PCR

(a) Denaturation (b) Annealing

(c) Primer extension (d) Non od these

17. PCR steps involves in correct way are

(a) Denaturation, Annealing and Primer extension

(b) Annealing, Denaturation and Primer extension

(c) Primer extension, annealing and Denaturation

(d) All of these

18. Basic requirement of PCR are

(a) DNA fragment and primer (b) dNTP and Taq polymerase

(c) Buffer solution and Mg++ (d) All of these

19. What is process of binding of primer in PCR steps

(a) Denaturation (b) Annealing

(c) Primer extension (d) All of these

20. What are temperature of PCR steps

(a) 94°C for 1 minute, 54 °C for 45 second and 72°C for 2 minutes

(b) 44°C for 1 minute, 34 °C for 45 second and 72°C for 2 minute

(c) 64°C for 1 minute, 54 °C for 45 second and 52°C for 2 minute

(d) 74°C for 1 minute, 54 °C for 45 second and 72°C for 2 minute

21. Complete steps of PCR cycle

(a) Denaturation (b) Annealing

(c) Primer extension (d) None of these

22. GC content of primer should be

(a) 45 to 55% (b) 60 to 70%

(c) 70 to 80% (d) None of these

23. Size of primer for PCR cycle

(a) 18 to 24 nucleotide (b) 30 to 35 nucleotide

(c) 35 to 40 nucleotide (d) 40 to 50 nucleotide

24. Formula of Melting temperature is

(a) Tm= [4x(G+C) +2x(A+T)] (b) Tm= [5 x (G+C) +4 x (A+T)]

(c) Tm= [2x(G+C) +4x(A+T)] (d) Tm= [3 x (G+C) +2 x (A+T)]

25. DNA template should be

(a) 1 kb (b) 2 kb

(c) 3 kb (d) 4 kb

26. Which of the following polymerase in which lack the proof-reading activity

(a) Pol I (b) Pol II

(c) Both (a) and (b) (d) Taq. Pol.

27. Which type of PCR technique that is used for amplifying of DNA from RNA template

(a) Real time PCR (b) Reverse transcription PCR

(c) Both (d) None of these

28. Take the time of each cycle of PCR

(a) 4 to 5 min. (b) 9 to 10 min.

(c) 10 to 20 min. (d) More than 20 min.

29. Klenow fragment is

(a) Pol I (b) Pol II

(c) Both (a) and (b) (d) Taq. Pol.

30. How many DNA duplexes are obtained after 4 cycles of PCR

(a) 15 (b)16

(c) 18 (d) 20

31. Microsatellite molecular marker is

(a) VNTR (b) SSR

(c) RAPD (d) Both (a) and (b)

32. Tissue culture technique was first practiced by

(a) Skoog (b) Haberlandt

(c) Halperin (d) None of these

33. Somaclones is

(a) Identical plants (b) Different plants

(c) Hybrid plants (d) None of these

34. Why virus free plants developed by meristem

(a) Meristem cell division faster than the rate of virus replication

(b) Virus affects to all parts of plant except meristem

(c) Both (a) and (b)

(d) None of these

35. Pomato is a

(a) Somatic hybridization (b) Anther culture

(c) Ovule culture (d) None of these

36. Callus formation occurs

(a) High auxin and low cytokinin

(b) low auxin and high cytokinin

(c) both are equal auxin and cytokinin

(d) None of these

37. Sorbitol, PVP and PEG is

(a) Permeable cryoprotectant (b) Non-Permeable cryoprotectant

(c) Both (a) and (b) (d) None of these

38. Protoplast is

(a) Artificial process

(b) Natural process

(c) Natural process but during grafting

(d) None of these

39. Carotenoids is a type of secondary metabolites

(a) Terpenoids (b) Steroids

(c) Alkaloids (d) None of these

40. First hydrated artificial seed was feveloped by

(a) Redenbergh (b) Mendel

(c) C. Linnaeus (d) None of these

41. Which agent used for the artificial seed

(a) Agar (b) Calcium alginate

(c) Carrageenins (d) All of these

42. Clone ward first given by

(a) Webber (b) Mendel

(c) Redenbergh (b) None of these

43. Organogenesis is

(a) Shoor formation (b) Root formation

(c) Both (a) and (b) (d) Only leaf

44. Which vitamin affect to the callus formation

(a) Vitamin A (b) Vitamin B2

(c) Vitamin C (d) All of these

45. which vitamin influence to shoot formation

(a) Vitamin A (b) Vitamin B_2

(c) Vitamin C (d) All of these

46. Carbohydrate metabolite occurs by

(a) Vitamin B_1 (b) Vitamin B_2

(c) Vitamin C (d) All of these

47. Which hormone promote to organogenesis

(a) Auxin (b) Cytokinin

(c) Ethylene (d) ABA

48. Which hormone inhibit to rooting and shooting

(a) GA_3 (b) Auxin

(c) Ethylene (d) ABA

49. First ovary culture

(a) San Noem 1976 (b) Melcher

(c) Morton (d) None of these

50. First Enzymatic method used for the isolation of protoplast

(a) Cocking (b) Skoog and Miller

(c) Hanning (d) White

[Part 3 - Answer Key]

Q.	Ans.	Q.	Ans.	Q.	Ans.	Q.	Ans.
1.	(b)	14.	(b)	27.	(b)	40.	(a)
2.	(a)	15.	(c)	28.	(a)	41.	(d)
3.	(b)	16.	(a)	29.	(a)	42.	(a)
4.	(a)	17.	(a)	30.	(b)	43.	(c)
5.	(b)	18.	(d)	31.	(d)	44.	(b)
6.	(b)	19.	(b)	32.	(b)	45.	(b)
7.	(b)	20.	(a)	33.	(a)	46.	(a)
8.	(a)	21.	(c)	34.	(a)	47.	(d)
9.	(d)	22.	(a)	35.	(a)	48.	(a)
10.	(d)	23.	(a)	36.	(c)	49.	(a)
11.	(c)	24.	(a)	37.	(b)	50.	(a)
12.	(a)	25.	(c)	38.	(c)		
13.	(b)	26.	(d)	39.	(a)		

Multiple Choice Questions

Part- 4

1. **Tissue culture is also known as**
   (a) Micropropagation (b) Anther culture
   (c) Pollen culture (d) SE
2. **Secondary metabolite is involved in**
   (a) Directly growth (b) Indirectly growth
   (c) Both (d) None of these
3. **Rubber is example of**
   (a) Alkaloids (b) Terpenoids
   (c) Phenylpropanoids (d) Steroids
4. **Barberine is**
   (a) Anti-inflammatory (b) Anti diabetic
   (c) Both (a) and (b) (d) None of these
5. **Barberine is obtained from**
   (a) Coptis japonica (b) Catharanthus roseus
   (c) Digitalis lanata (d) None of these
6. **Secondary metabolite reduced by**
   (a) 2, 4 D (b) NH_4
   (c) Cytokinin (d) All of these
7. **Most suitable stage for embryo excision for culture is**
   (a) Globular stage (b) Heart stage
   (c) Both (a) and (b) (d) None of these
8. **Totipotency concept given by**
   (a) Hanning (b) Skoog
   (c) Steward (d) G. Haberlandt 1902

9. Discovery of somatic embryogenesis

(a) Steward (b) Skoog and Miller

(c) Hanning (d) None of these

10. Development of Pomato from fusion of potato and tomato together by

(a) Melcher (b) San Noem

(c) Morton (d) Power

11. First haploid plant developed from microspore culture

(a) Guha and Maheshwari 1964 (b) Power

(c) Morton (d) None of these

12. Developing from unusual points of origin, such as shoot or roots arising from a leaf

(a) Adventitious (b) Meristem

(c) Suspension culture (d) Culture

13. Meristematic is

(a) High mitotic activity (b) Low mitotic activity

(c) Both (d) None of these

14. Organogenesis is type of

(a) Direct (without callus) (b) Indirect (with callus)

(c) Both (a) and (b) (d) None of these

15. Regulation of organogenesis by

(a) Auxin (b) Cytokinin

(c) Both (a) and (b) (d) None of these

16. Somatic embryogenesis was first demonstrated by

(a) Reinert and Steward (b) Hanning

(c) Skoog (d) Steward

17. Acclimatization is

(a) Plant transfer from Labe to the field

(b) Shoot formation

(c) Root formation

(d) Lant from field to Labe transfer

18. Virus free plant developed by

(a) Morel and Martin 1952 in dahlia

(b) Skoog in dahlia

(c) Morel in dahlia

(d) Hanning in dahlia

19. Best explant for virus free plants is

(a) Shoot tip (b) Root cap

(c) Both (d) None of these

20. Size of shoot tip should be

(a) 0.1mm (b) 0.1 to 1.00 mm

(c) More than 1.0 mm (d) 5.0 mm

21. Morphine secondary metabolite is obtained from

(a) Papaver seminiferum (b) Morinda citrifolia

(c) Both (d) None of these

22. Shikonine is a

(a) Secondary metabolite (b) Biochemical product

(c) Both (a) and (b) (d) None of these

23. Auxin does not require for

(a) Anther culture (b) Pollen culture

(c) Embryo culture (d) All of these

24. Media suitable for pollen / anther culture

(a) N_6 Media (b) MS Media

(c) LS Media (d) All of these

25. Caulogenesis is

(a) Process of root formation (b) Process of shoot formation

(c) Both (a) and (b) (d) None of these

26. Rhizogenesis is

(a) Process of root formation (b) Process of shoot formation

(c) Both (a) and (b) (d) None of these

27. Tryptophane is precursor of

(a) GA (b) Auxin

(c) Cytokinin (d) ABA

28. TIR 1, ABP 1 are receptors of

(a) GA (b) Auxin

(c) Cytokinin (d) ABA

29. Adenine is precursor of

(a) ABA (b) Cytokinin

(c) Auxin (d) GA

30. Methionine is precursor of

(a) Auxin (b) Ethylene

(c) GA (d) None of these

31. ETR 1, ERS, EIN 4 are receptors of

(a) Auxin (b) Ethylene

(c) GA (d) None of these

32. Lenoleic acid is precursos of

(a) Brassinosterods (b) Auxin

(c) Jasmonic acid (d) None of these

33. Lenolenic acid is precursor of

(a) Brassinosterods (b) Auxin

(c) Jasmonic acid (d) None of these

34. Acetyl CoA is precursor of

(a) GA (b) Cytokinin

(c) Auxin (d) ABA

35. GID 1 is receptor of

(a) GA (b) Cytokinin

(c) Auxin (d) ABA

36. Xanthophyll is precursor of

(a) GA (b) Cytokinin

(c) Auxin (d) ABA

37. Immature seed are very rich in

(a) GA (b) Cytokinin

(c) Auxin (d) ABA

38. Embryonic clump in somatic embryogenesis is developed

(a) After callus

(b) Before callus

(c) Before mature somatic embryo

(d) Both (a) and (c)

39. Embryo developed by direct explant called as

(a) Indirect embryogenesis (b) Direct embryogenesis

(c) Both (a) and (b) (d) None of these

40. Development of bipolar structure from non-zygotic somatic cells called as

(a) Organogenesis (b) Somatic embryogenesis
(c) Haploid culture (d) None of these

41. Organogenesis is

(a) Monopolar (b) Bipolar
(c) both (a) and (b) (d) None of these

42. Somatic embryogenesis is

(a) Monopolar (b) Bipolar
(c) Both (a) and (b) (d) None of these

43. Which amino acid is mostly used for the somatic embryogenesis

(a) Glycine (b) Glutamine
(c) L-Glutamine (d) Alanine

44. Which chemical used for the double haploid

(a) Colchicine treatment (b) Heat treatment
(c) Cold treatment (d) All of these

45. Plant cell without cell wall is known as

(a) Protoplast (b) Protoplasm
(c) Cytoplasm (d) All of these

46. Pectinase is source of

(a) A. niger (b) A. japonica
(c) T. viridae (d) None of these

47. Protoplast regenerated cell wall within

(a) 6 day (b) 2 to 4 day
(c) 10 day (d) 30 day

48. Why addition to "Potassium Dextran Sulfate" during protoplast isolation

(a) For hydrolysis (b) Protect to the protoplast from hydrolysis
(c) Both (a) and (b) (d) None of these

49. PEG mediated protoplast fusion is

(a) Selective fusion (b) Non selective method
(c) Both (a) and (b) (d) None of these

50. Embryo rescue is

(a) Embryo culture for production of interspecific hybrid

(b) Embryo culture for production of intraspecific hybrid

(c) Both (a) and (b)

(d) All of the above

[Part 4 - Answer Key]

Q.	*Ans.*	*Q.*	*Ans.*	*Q.*	*Ans.*	*Q.*	*Ans.*
1.	(a)	**14.**	(c)	**27.**	(b)	**40.**	(b)
2.	(b)	**15.**	(c)	**28.**	(b)	**41.**	(a)
3.	(b)	**16.**	(a)	**29.**	(b)	**42.**	(b)
4.	(c)	**17.**	(a)	**30.**	(b)	**43.**	(c)
5.	(a)	**18.**	(a)	**31.**	(b)	**44.**	(a)
6.	(d)	**19.**	(a)	**32.**	(a)	**45.**	(a)
7.	(b)	**20.**	(b)	**33.**	(c)	**46.**	(a)
8.	(d)	**21.**	(a)	**34.**	(a)	**47.**	(b)
9.	(a)	**22.**	(c)	**35.**	(a)	**48.**	(b)
10.	(a)	**23.**	(c)	**36.**	(d)	**49.**	(b)
11.	(a)	**24.**	(a)	**37.**	(a)	**50.**	(a)
12.	(a)	**25.**	(b)	**38.**	(d)		
13.	(a)	**26.**	(a)	**39.**	(b)		

Fill in the blanks

Part- 1

Fill in the Blanks

1. Fusion of one protoplast with enucleated protoplast called as...............
2. Micropropagation is also called as ....
3. Tryptophane is precursor of ....
4. Somaclonal variation given by .... In 1981
5. Sigmoid curve occurs in ....
6. First plant hormone is ....
7. MS media was developed by ....
8. Who was scientist developed First haploid plant by microspore culture ....
9. First plant tissue culture media is................
10. Autoclave sterilization at ......°C, ...psi (pounds per square inch) for ....min.
11. .... % Ethanol used for surface sterilization
12. .... used for sterilization of explant
13. Agar is purified from ....
14. Solidifying agent is .... used in plant tissue culture media
15. Endosperm is ....

Part-1 Answer Key

1.	Cybrid	2.	Clonal propagation
3.	Auxin	4.	Larkin and Scowkraft (1981)
5.	Batch culture	6.	Auxine (IBA)
7.	Murashige and Skoog	8.	Guha and Maheshwari 1964

9. M S Media
10. 120, 15, 20
11. 70
12. Mercuric clioride
13. Red algae
14. Agar (0.6% to 0.8%)
15. Triploid (3n)

Fill in the blanks

Part-2

Fill in the Blanks

1. Meristem is obtained virus free plant in dahlia by ....and .... (1952)
2. Best explant for virus free plants............
3. Virus free plant developed by ....plant
4. Secondary metabolite developed by ....culture
5. Culture show sigmoid growth pattern........
6. Kinetin is form of ....plant hormone
7. Most suitable stage for embryo excision for culture...........
8. embryo rescue is a type of ....
9. DAMA is variety of ....
10. Most suitable media for haploid...........
11. Plant cell without cell wall is.......
12. Protoplast fused together through ....
13. Calliclone obtained from.............
14. Pusa jai kisan is a variety of ....developed by........
15. DMSO is ....

Part-2 Answer Key

1.	Morel and Martin	2.	Apical meristem
3.	Meristem culture	4.	Cell suspension culture
5.	Batch culture	6.	Cytokinin
7.	Heart shaped stage	8.	Immature embryo
9.	Rice	10.	NN Media

11. Protoplast
12. PEG (polyethylene glycol)
13. Callus
14. Mustard , somaclonal variation
15. Cryoprotectant agent

Fill in the blanks

Part-3

Fill in the Blanks

1. Callus is ....
2. If auxin and cytokinin is intermediated: it lead to .... formation.
3. if auxin/cytokinin is high: it promote..........
4. If auxin / cytokinin is low: it promote.........
5. Optimum pH should be .... for plant tissue culture.
6. Agar (gelling agent) should be ....
7. Calcium alginate is used for the production of .... Culture.
8. Organogenesis is type of .... structure
9. Somatic embryogenesis is a .... structure
10. Hyperhydricity is also known as ....
11. Morphine, Codein, Quinine, Nicotine are.............
12. Parabolic / Hyperbolic growth pattern occurs in ....culture.
13. Glycine is a............... amino acid
14. Requirement of micro nutrient in culture media ....
15. Requirement of macro nutrient in culture media ....

Part-3 Answer Key

1. Un organized mass of cell
2. Callus
3. Root
4. Shoot
5. 5.0 to 6.0
6. 0.6 % to 0.8%
7. Haploid
8. Monopolar
9. Bipolar
10. Vitrification

11. Secondary metabolite
12. Continuous
13. Neutral
14. < 0.5 mM
15. > 0.5 mM

Fill in the blanks

Part-4

Fill in the Blanks

1. Banana is a .... crop
2. Preservation of the germplasm under natural environment ....
3. Preservation in artificial condition ....
4. Temperature of liquid nitrogen ....
5. In plant tissue culture, Laminar Air Flow is used for ....
6. Hot air oven is used for..........
7. Centrifuge is used for the ....
8. Gel electrophoresis is used for the ....
9. Water bath is used to ....
10. pipette is used for ....
11. PCR is laboratory equipment which is used for ....
12. Plant transfer from lab to pots or field called as......
13. CTAB is a method of ....
14. SSR stand for ....
15. RFLP Stand for ....

Part -4 Answer Key

1. Triploid
2. In-situ
3. Ex-situ
4. -196°C
5. Inoculation of explant in culture media
6. Dissolve the solution
7. Separate pellet from the solution
8. Isolation of DNA or RNA
9. Incubate sample at constant temperature
10. Transport a measure volume of liquid in small amount

11. Application of DNA
12. Hardening
13. DNA isolation
14. Simple Sequence Repeats
15. Restriction Fragment Length Polymorphism

www.ingramcontent.com/pod-product-compliance
Ingram Content Group UK Ltd.
Pitfield, Milton Keynes, MK11 3LW, UK
UKHW021951270726
14060UKWH00002B/467

9 789359 194264